A Guide to
Distribution Theory and
Fourier Transforms

Robert S Strichartz

Cornell University, USA

A Guide to
Distribution Theory and
Fourier Transforms

World Scientific

NEW JERSEY • LONDON • SINGAPORE • BEIJING • SHANGHAI • HONG KONG • TAIPEI • CHENNAI

Published by

World Scientific Publishing Co. Pte. Ltd.

5 Toh Tuck Link, Singapore 596224

USA office: 27 Warren Street, Suite 401-402, Hackensack, NJ 07601

UK office: 57 Shelton Street, Covent Garden, London WC2H 9HE

Library of Congress Cataloging-in-Publication Data
Strichartz, Robert S.
 A guide to distribution theory and Fourier transforms / Robert S. Strichartz.
  p. cm.
 Includes bibliographical references and index.
 ISBN-13 978-981-238-421-8
 ISBN-10 981-238-421-9
 ISBN-13 978-981-238-430-0 (pbk)
 ISBN-10 981-238-430-8 (pbk)
 1. Theory of distributions (Functional analysis). 2. Fourier analysis. I. Title.
 QA324 .S77 2003
 515'.782--dc20

        2004540483

British Library Cataloguing-in-Publication Data
A catalogue record for this book is available from the British Library.

First edition in 1994 by CRC Press
Copyright © 1994 CRC Press

Reprinted 2008, 2015

Preface

Distribution theory was one of the two great revolutions in mathematical analysis in the 20th century. It can be thought of as the completion of differential calculus, just as the other great revolution, measure theory, (or Lebesgue integration theory), can be thought of as the completion of integral calculus. There are many parallels between the two revolutions. Both were created by young, highly individualistic French mathematicians (Henri Lebesgue and Laurent Schwartz). Both were rapidly assimilated by the mathematical community, and opened up new worlds of mathematical development. Both forced a complete rethinking of all mathematical analysis that had come before, and basically altered the nature of the questions that mathematical analysts asked. (This is the reason I feel justified in using the word "revolution" to describe them). But there are also differences. When Lebesgue introduced measure theory (circa 1903), it almost came like a bolt from the blue. Although the older integration theory of Riemann was incomplete—there were many functions that did not have integrals—it was almost impossible to detect this incompleteness from within, because the non-integrable functions really appeared to have no well defined integral. As evidence that the mathematical community felt perfectly comfortable with Riemann's integration theory, one can look at Hilbert's famous list (dating to 1900) of 23 unsolved problems that he thought would shape the direction of mathematical research in the 20th century. Nowhere is there a hint that completing integration theory was a worthwhile goal. On the other hand, a number of his problems do foreshadow the developments that led to distribution theory (circa 1945). When Laurent Schwartz came out with his theory, he addressed problems that were of current interest, and he was able to replace a number of more complicated theories that had been developed earlier in an attempt to deal with the same issues.

From the point of view of this work, the most important difference is that in retrospect, measure theory still looks hard, but distribution theory looks easy. Because it is relatively easy, distribution theory should be accessible to a wide audience, including users of mathematics and mathematicians who specialize in other fields. The techniques of distribution theory can be used, confidently and effectively—just like the techniques of calculus are used—without a complete knowledge of the formal mathematical foundations of the subject. The aim of this book is thus very similar to the aim of a typical calculus textbook: to explain the techniques of the theory with precision, to provide an intuitive discussion of the ideas that underline the techniques, and to offer a selection of problems applying the techniques.

Because the Lebesgue theory of integration preceded distribution theory historically, and is required for the rigorous mathematical development of the theory, it might be thought that a knowledge of the Lebesgue theory would have to be a prerequisite for studying distribution theory. I do not believe that this is true, and I hope this book makes a good case for my point of view. When you see an integral sign in this book, you are free to interpret it in the sense of any integration theory you have learned. If you have studied the Lebesgue theory in any form, then of course think of the integrals as Lebesgue integrals. But if not, don't worry about it. Let the integral mean what you think it means.

Distribution theory is a powerful tool, but it becomes an even more powerful tool when it works in conjunction with the theory of Fourier transforms. One of the main areas of applications is to the theory of partial differential equations. These three theories form the main themes of this book. The first two chapters motivate and introduce the basic concepts and computational techniques of distribution theory. Chapters three and four do the same for Fourier transforms. Chapter five gives some important and substantial applications to particular partial differential equations that arise in mathematical physics. These five chapters, part I of the book, were written with the goal of getting to the point as quickly as possible. They have been used as a text for a portion of a course in applied mathematics at Cornell University for more than 10 years.

The last three chapters, part II of the book, return to the three themes in greater detail, filling in topics that were left aside in the rapid development of part I, but which are of great interest in and of themselves, and point toward further applications. Chapter six returns to distribution theory, explaining the notion of continuity, and giving the important structure theorems. Chapter seven covers Fourier analysis. In addition to standard material, I have included some topics of recent origin, such as quasicrystals and wavelets. Finally, Chapter eight returns to partial differential equations, giving an introduction to the modern theory of general linear equations. Here the reader will meet Sobolev spaces, a priori estimates, equations of elliptic and hyperbolic type, pseudodifferential operators, wave front sets, and the ideology known as microlocal analysis. Part II was written for this book, and deals with somewhat more abstract material. It was not designed for use as a textbook, but more to satisfy the curiosity of those readers of part I who want to learn in greater depth about the material. I also hope it will serve as an appetizer for readers who will go on to study these topics in greater detail.

The prerequisites for reading this book are multidimensional calculus and an introduction to complex analysis. A reader who has not seen any complex analysis will be able to get something out of this book, but will have to accept that there will be a few mystifying passages. A solid background in multi-dimensional calculus is essential, however, especially in part II.

Recently, when I was shopping at one of my favorite markets, I met a graduate of Cornell (who had not been in any of my courses). He asked me what I was doing, and when I said I was writing this book, he asked sarcastically "do you guys enjoy writing them as much as we enjoy reading them?" I don't know what other books he had in mind, but in this case I can say quite honestly that I very much enjoyed writing it. I hope you enjoy reading it.

Acknowledgments:

I am grateful to John Hubbard, Steve Krantz and Wayne Yuhasz for encouraging me to write this book, and to June Meyermann for the excellent job of typesetting the book in LATEX.

Ithaca, NY
October 1993

Contents

Preface   **v**

1 What are Distributions?   **1**
   1.1 Generalized functions and test functions   1
   1.2 Examples of distributions   5
   1.3 What good are distributions?   8
   1.4 Problems   10

2 The Calculus of Distributions   **12**
   2.1 Functions as distributions   12
   2.2 Operations on distributions   14
   2.3 Adjoint identities   18
   2.4 Consistency of derivatives   20
   2.5 Distributional solutions of differential equations   22
   2.6 Problems   25

3 Fourier Transforms   **28**
   3.1 From Fourier series to Fourier integrals   28
   3.2 The Schwartz class S   31
   3.3 Properties of the Fourier transform on S   32
   3.4 The Fourier inversion formula on S   38
   3.5 The Fourier transform of a Gaussian   41
   3.6 Problems   43

4 Fourier Transforms of Tempered Distributions   **46**
   4.1 The definitions   46
   4.2 Examples   49
   4.3 Convolutions with tempered distributions   55
   4.4 Problems   57

5 Solving Partial Differential Equations   **60**
   5.1   The Laplace equation   60
   5.2   The heat equation   64
   5.3   The wave equation   67
   5.4   Schrödinger's equation and quantum mechanics   72
   5.5   Problems   73

6 The Structure of Distributions   **78**
   6.1   The support of a distribution   78
   6.2   Structure theorems   82
   6.3   Distributions with point support   85
   6.4   Positive distributions   88
   6.5   Continuity of distribution   91
   6.6   Approximation by test functions   98
   6.7   Local theory of distributions 101
   6.8   Distributions on spheres 103
   6.9   Problems 108

7 Fourier Analysis   **113**
   7.1   The Riemann-Lebesgue lemma 113
   7.2   Paley-Wiener theorems 119
   7.3   The Poisson summation formula 125
   7.4   Probability measures and positive definite functions 130
   7.5   The Heisenberg uncertainty principle 134
   7.6   Hermite functions 139
   7.7   Radial Fourier transforms and Bessel functions 143
   7.8   Haar functions and wavelets 149
   7.9   Problems 157

8 Sobolev Theory and Microlocal Analysis   **162**
   8.1   Sobolev inequalities 162
   8.2   Sobolev spaces 172
   8.3   Elliptic partial differential equations (constant coefficients) 176
   8.4   Pseudodifferential operators 185
   8.5   Hyperbolic operators 191
   8.6   The wave front set 200
   8.7   Microlocal analysis of singularities 209
   8.8   Problems 214

Suggestions for Further Reading   **219**

Index   **221**

Chapter 1

What are Distributions?

1.1 Generalized functions and test functions

You have often been asked to consider a function $f(x)$ as representing the value of a physical variable at a particular point x in space (or space-time). But is this a realistic thing to do? Let us borrow a perspective from quantum theory and ask: What can you measure?

Suppose that $f(x)$ represents temperature at a point x in a room (or if you prefer let $f(x,t)$ be temperature at point x and time t.) You can measure temperature with a thermometer, placing the bulb of the thermometer at the point x. Unlike the point, the bulb of the thermometer has a nonzero size, so what you measure is more an average temperature over a small region of space (again if you think of temperature as varying with time also, then you are also averaging over a small time interval preceeding the time t when you actually read the thermometer). Now there is no reason to believe that the average is "fair" or "unbiased." In mathematical terms, a thermometer measures

$$\int f(x)\varphi(x)\,dx$$

where $\varphi(x)$ depends on the nature of the thermometer and where you place it—$\varphi(x)$ will tend to be "concentrated" near the location of the thermometer bulb and will be nearly zero once you are sufficiently far away from the bulb. To say this is an "average" is to require

$$\varphi(x) \geq 0 \text{ everywhere, and}$$

$$\int \varphi(x)\,dx = 1 \text{ (the integral is taken over all space).}$$

1

However, do not let these conditions distract you. With two thermometers you can measure

$$\int f(x)\varphi_1(x)\,dx \qquad \text{and} \qquad \int f(x)\varphi_2(x)\,dx$$

and by subtracting you can deduce the value of $\int f(x)[\varphi_1(x) - \varphi_2(x)]\,dx$. Note that $\varphi_1(x) - \varphi_2(x)$ is no longer nonnegative. By doing more arithmetic you can even compute $\int f(x)(a_1\varphi_1(x) - a_2\varphi_2(x))\,dx$ for constants a_1 and a_2, and $a_1\varphi_1(x) - a_2\varphi_2(x)$ may have any finite value for its integral.

The above discussion is meant to convince you that it is often more meaningful physically to discuss quantities like $\int f(x)\varphi(x)\,dx$ than the value of f at a particular point x. The secret of successful mathematics is to eliminate all unnecessary and irrelevant information—a mathematician would not ask what color is the thermometer (neither would an engineer, I hope). Since we have decided that the value of f at x is essentially impossible to measure, let's stop requiring our functions to have a value at x. That means we are considering a larger class of objects. Call them *generalized functions*. What we will require of a generalized function is that something akin to $\int f(x)\varphi(x)\,dx$ exist for a suitable choice of averaging functions φ (call them *test functions*). Let's write $\langle f, \varphi \rangle$ for this something. It should be a real number (or a complex number if we wish to consider complex-valued test functions and generalized functions). What other properties do we want? Let's recall some arithmetic we did before, namely

$$a_1 \int f(x)\varphi_1(x)\,dx - a_2 \int f(x)\varphi_2(x)\,dx = \int f(x)(a_1\varphi_1(x) - a_2\varphi_2(x))\,dx.$$

We want to be able to do the same sort of thing with generalized functions, so we should require $a_1\langle f, \varphi_1 \rangle - a_2\langle f, \varphi_2 \rangle = \langle f, a_1\varphi_1 - a_2\varphi_2 \rangle$. This property is called *linearity*. The minus sign is a bit silly, since we can get rid of it by replacing a_2 by $-a_2$. Doing this we obtain the condition

$$a_1\langle f, \varphi_1 \rangle + a_2\langle f, \varphi_2 \rangle = \langle f, a_1\varphi_1 - a_2\varphi_2 \rangle.$$

Notice we have tacitly assumed that if φ_1, φ_2 are test functions then $a_1\varphi_1 + a_2\varphi_2$ is also a test function. I hope these conditions look familiar to you—if not, please read the introductory chapter of any book on linear algebra.

You have almost seen the entire definition of *generalized functions*. All you are lacking is a description of what constitutes a test function and one technical hypothesis of continuity. Do not worry about continuity—it will always be satisfied by anything you can *construct* (wise-guys who like using the axiom of choice will have to worry about it, along with wolves under the bed, etc).

So, now, what are the test functions? There are actually many possible choices for the collection of test functions, leading to many different theories of generalized functions. I will describe the space called $\mathcal{D}$, leading to the theory of distributions. Later we will meet other spaces of test functions.

The underlying point set will be an n-variable space $\mathbb{R}^n$ (points x stand for $x = (x_1, \ldots, x_n)$) or even a subset $\Omega \subset \mathbb{R}^n$ that is open. Recall that this means every point $x \in \Omega$ is surrounded by a ball $\{y : |x - y| < \epsilon\}$ contained in Ω, where ϵ depends on x, and

$$|x - y| = \sqrt{(x_1 - y_1)^2 + \cdots + (x_n - y_n)^2}.$$

Of course $\Omega = \mathbb{R}^n$ is open, as is every open ball $\{y : |x - y| < r\}$. Intuitively, an open set is just a union of open balls.

The class of test functions $\mathcal{D}(\Omega)$ consists of *all functions $\varphi(x)$ defined in Ω, vanishing outside a bounded subset of Ω that stays away from the boundary of Ω, and such that all partial derivatives of all orders of φ are continuous.*

For example, if $n = 1, \Omega = \{0 < x < 1\}$, then

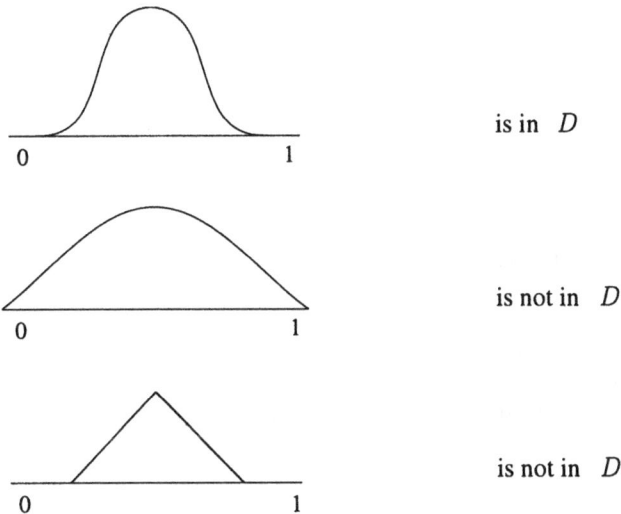

is in D

is not in D

is not in D

Figure 1.1

The second example fails because it does not vanish near the boundary point 0, and the third example fails because it is not differentiable at three points. To actually write down a formula for a function in $\mathcal{D}$ is more difficult. Notice that no analytic function (other than $\varphi \equiv 0$) can be in $\mathcal{D}$ because of the vanishing requirement. Thus any formula for φ must be given "in pieces." For example, in $\mathbb{R}^1$

$$\psi(x) = \begin{cases} e^{-1/x^2} & x > 0 \\ 0 & x \le 0 \end{cases}$$

has continuous derivatives of all order:

$$\left(\frac{d}{dx}\right)^k e^{-1/x^2} = \frac{\text{polynomial in } x}{\text{polynomial in } x} e^{-1/x^2}$$

and as $x \to 0$ this approaches zero since the zero of e^{-1/x^2} beats out the pole of $\frac{1}{\text{polynomial in } x}$. Thus $\varphi(x) = \psi(x)\psi(1-x)$ has continuous derivatives of all orders (we abbreviate this by saying φ is C^∞) and vanishes outside $0 < x < 1$, so $\varphi \in \mathcal{D}(\mathbb{R}^1)$; in fact, $\varphi \in \mathcal{D}(a < x < b)$ provided $a < 0$ and $b > 1$ (why not $a \le 0$ and $b \ge 1$?). Once you have one example you can manufacture more by

1. moving it about $\varphi(x + x_0)$

2. changing vertical scale $a\varphi(x)$

3. changing horizontal scale $\varphi(ax)$

4. taking linear combinations $a_1\varphi_1(x) + a_2\varphi_2(x)$ if $\varphi_1, \varphi_2 \in \mathcal{D}$.

5. taking products $\varphi_1(x_1,\ldots,x_n) = \varphi(x_1)\varphi(x_2)\ldots\varphi(x_n)$ to obtain examples in higher dimensions.

Exercise: Draw the pictures associated with operations 1–4.

These considerations should convince you that you can make a test function in $\mathcal{D}$ do anything you can draw a picture of a smooth function doing. You can make it take on prescribed values at a finite set of points, make it vanish on any open set and even take on a constant value (say 1) on a bounded open set away from the boundary (this requires a little more work).

O.K. That is what $\mathcal{D}(\Omega)$ is. We can now define the class of distributions on Ω, denoted $\mathcal{D}'(\Omega)$, to be all continuous *linear functionals* on $\mathcal{D}(\Omega)$. By *functional* I mean a real (or complex) -valued function on $\mathcal{D}(\Omega)$, written $\langle f, \varphi \rangle$. By *linear* I mean it satisfies the identity

$$a_1\langle f, \varphi_1 \rangle + a_2\langle f, \varphi_2 \rangle = \langle f, a_1\varphi_1 + a_2\varphi_2 \rangle.$$

(Yes, Virginia, $a_1\varphi_1 + a_2\varphi_2$ is in $\mathcal{D}(\Omega)$ if φ_1 and φ_2 are in $\mathcal{D}(\Omega)$.) By *continuous* I mean that if φ_1 is close enough to φ then $\langle f, \varphi_1 \rangle$ is close to $\langle f, \varphi \rangle$—the exact definition can wait until later. Continuity has an intuitive physical interpretation—you want to be sure that different thermometers give approximately the same reading provided you control the manufacturing process adequately. Put another way, when you repeat an experiment you do not want to get a different answer because small experimental errors get magnified. Now, whereas discontinuous *functions* abound, *linear functionals* all tend to be continuous. This happy fact deserves a bit of explanation. Fix φ and φ_1 and call the difference $\varphi_1 - \varphi = \varphi_2$. Then $\varphi_1 = \varphi + \varphi_2$. Now perhaps $\langle f, \varphi \rangle$ and $\langle f, \varphi_1 \rangle$ are far apart. So what? Move φ_1 closer to φ by considering $\varphi + t\varphi_2$ and let t get small. Then $\langle f, \varphi + t\varphi_2 \rangle = \langle f, \varphi \rangle + t \langle f, \varphi_2 \rangle$ by linearity, and as t gets small this gets close to $\langle f, \varphi \rangle$. This does *not* constitute a proof of continuity, since the definition requires more "uniformity," but it should indicate that a certain amount of continuity is built into linearity. At any rate, all linear functionals on $\mathcal{D}(\Omega)$ you will ever encounter will be continuous.

1.2 Examples of distributions

Now for some examples. Any function gives rise to a distribution by setting $\langle f, \varphi \rangle = \int_\Omega f(x)\varphi(x)\,dx$, at least if the integral can be defined. This is certainly true if f is continuous, but actually more general functions will work. Depending on what theory of integration you are using, you may make f discontinuous and even unbounded, provided the improper integral converges *absolutely*. For instance $\int_{|x|\le r} |x|^{-t}\,dx$ converges for $t < n$ (in n dimensions) and diverges for $t \ge n$. Thus the function $|x|^{-t}$ for $t < n$ gives rise to the distribution in $\mathcal{D}'(\mathbb{R}^n) \langle f, \varphi \rangle = \int_{\mathbb{R}^n} \varphi(x)|x|^{-t}\,dx$ (the actual range of integration is bounded since φ vanishes outside a bounded set).

A different sort of example is the Dirac δ-function: $\langle \delta, \varphi \rangle = \varphi(0)$. In this case we have to check the linearity property, but it is trivial to verify:

$$a_1 \langle \delta, \varphi_1 \rangle + a_2 \langle \delta, \varphi_2 \rangle = a_1\varphi_1(0) + a_2\varphi_2(0) = \langle \delta, a_1\varphi_1 + a_2\varphi_2 \rangle.$$

An even wilder example is δ' (now in $\mathcal{D}'(\mathbb{R}^1)$) given by $\langle \delta', \varphi \rangle = -\varphi'(0)$.

Exercise: Verify linearity.

These examples demand a closer look, some pictures, and an explanation of the minus sign. Consider any function $f_k(x)$ (for simplicity we work in one dimension) that satisfies the conditions

1. $f_k(x) = 0$ unless $|x| \le \frac{1}{k}$

   2. $\int_{-1/k}^{1/k} f_k(x)\,dx = 1$.

The simplest examples are

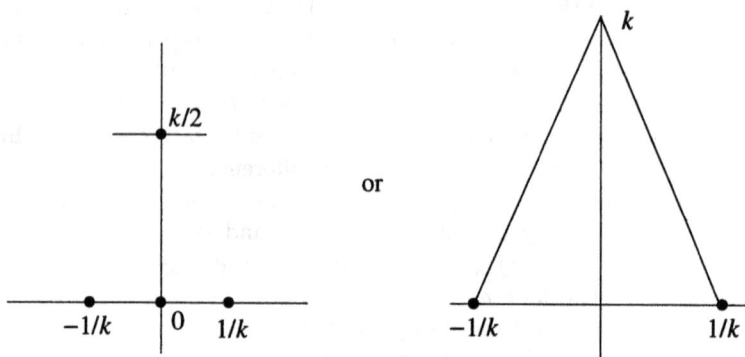

or

Figure 1.2

but we may want to take f_k smoother, even in $\mathcal{D}$

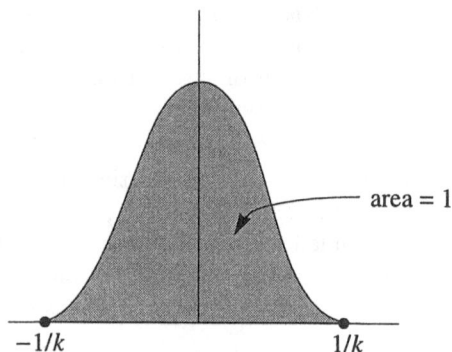

area = 1

Figure 1.3

Now the distribution

$$\langle f_k, \varphi \rangle = \int f_k(x)\varphi(x)\,dx$$

is an average of φ near zero, so that if φ does not vary much in $-1/k \le x \le 1/k$, it is close to $\varphi(0)$. Certainly in the limit as $k \to \infty$ we get

$\langle f_k, \varphi \rangle \rightarrow \varphi(0) = \langle \delta, \varphi \rangle$. Thus we may think of δ as $\lim_{k \to \infty} f_k$. (Of course, pointwise

$$\lim_{k \to \infty} f_k(x) = \begin{cases} 0 & \text{if } x \neq 0 \\ +\infty & \text{if } x = 0 \end{cases}$$

for suitable choice of f_k, but this is nonsense, showing the futility of pointwise thinking.)

Now suppose we first differentiate f_k and then let $k \rightarrow \infty$?

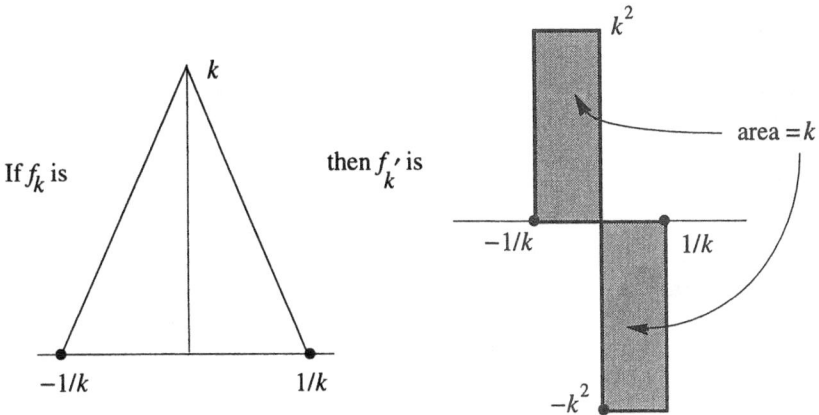

Figure 1.4

and

$$\langle f'_k, \varphi \rangle \approx \frac{\varphi\left(-\frac{1}{2k}\right) - \varphi\left(\frac{1}{2k}\right)}{1/k}$$

(the points $-\frac{1}{2k}$ and $\frac{1}{2k}$ are the midpoints of the intervals and the factor $(1/k)^{-1} = k$ is the area) which approaches $-\varphi'(0)$ as $k \rightarrow \infty$. We obtain the same answer formally by integrating by parts:

$$\int f'_k(x)\varphi(x)\,dx = -\int f_k(x)\varphi'(x)\,dx = -\langle f_k, \varphi' \rangle \rightarrow -\varphi'(0)$$

as $k \rightarrow \infty$. Here we might assume that f_k is continuously differentiable. Note that there are no boundary terms in the integration-by-parts formula because φ vanishes for large values of x. Thus if $f_k \rightarrow \delta$ then $f'_k \rightarrow \delta'$, which justifies the notation of δ' as the derivative of δ.

1.3   What good are distributions?

Enough examples for now. We will have more later. Let us pause to
consider the following question: Why this particular (perhaps peculiar)
choice of test functions $\mathcal{D}$? The answer is not easy. In fact, the theory
of generalized functions was in use for almost twenty years before Laurent
Schwartz proposed the definition of $\mathcal{D}$. So it is not possible to use physical
or intuitive grounds to say $\mathcal{D}$ is the only "natural" class of test functions.
However, it does yield an elegant and useful theory—so that after the fact
we may thank Laurant Schwartz for his brilliant insight. You can get some
feel for what is going on if you observe that the smoother you require the
test functions φ to be, the "rougher" you can allow the generalized functions
f to be. To define $\langle \delta, \varphi \rangle$, φ must be at least continuous, and to define $\langle \delta', \varphi \rangle$,
you must require φ to be differentiable. Later I will show you how to define
derivatives for any distribution—the key point in the definition will be the
ability to differentiate the test functions.

The requirement that test functions in $\mathcal{D}$ vanish outside a bounded set
and near the boundary of Ω is less crucial. It allows distributions to "grow
arbitrarily rapidly" as you approach the boundary (or infinity). Later we
will consider a *smaller* class of distributions, called *tempered distributions*,
which cannot grow as rapidly at infinity, by considering a *larger* class of
test functions that have weaker vanishing properties.

Another question you should be asking at this point is: What good are
distributions? Let me give a hint of one answer. Differential equations are
used to construct models of reality. Sometimes the reality we are model-
ing suggests that some solutions of the differential equation need not be
differentiable! For example, the "vibrating string" equation

$$\frac{\partial^2 u(x,t)}{\partial t^2} = k^2 \frac{\partial^2 u(x,t)}{\partial x^2}$$

has a solution $u(x,t) = f(x - kt)$ for any function of one variable f, which
has the physical interpretation of a "traveling wave" with "shape" $f(x)$
moving at velocity k.

There is no physical reason for the "shape" to be differentiable, but if it
is not, the differential equation is not satisfied at some points. But we do
not want to throw away physically meaningful solutions because of techni-
calities. You might be tempted therefore to think that if a function satisfies
a differential equation except for some points where it is not differentiable,
it should be admitted as a solution. The next example shows that such a
simplistic idea does not work.

Laplace's equation $\Delta u = 0$ where Δ (called the *Laplacian* and some-

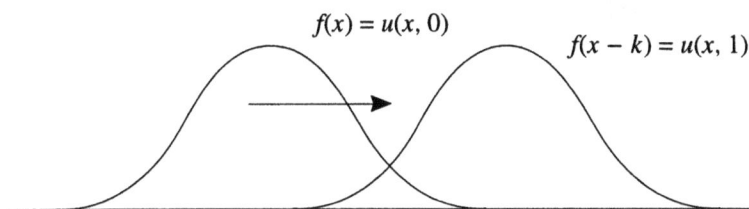

$f(x) = u(x, 0)$

$f(x - k) = u(x, 1)$

Figure 1.5

times written ∇^2) is defined by

$$\Delta u = \frac{\partial^2 u}{\partial x_1^2} + \frac{\partial^2 u}{\partial x_2^2} \quad \text{in } \mathbb{R}^2$$

and

$$\frac{\partial^2 u}{\partial x_1^2} + \frac{\partial^2 u}{\partial x_2^2} + \frac{\partial^2 u}{\partial x_3^2} \quad \text{in } \mathbb{R}^3.$$

A solution to Laplace's equation has the physical interpretation of a potential in a region without charge. A straightforward (albeit long) calculation shows that $u(x_1, x_2) = \log(x_1^2 + x_2^2)$ and $u(x_1, x_2, x_3) = (x_1^2 + x_2^2 + x_3^2)^{-1/2}$ are solutions at every point except the origin, where they fail to be differentiable. But they must be rejected on physical grounds because potentials away from charges must be continuous (even smooth), while these functions have poles.

The moral of the story is that distribtution theory allows you to distinguish between these two cases. In the first case $u(x, t) = f(x - kt)$, as a distribution in $\mathbb{R}^2$, satisfies the vibrating string equation. But $u(x_1, x_2) = \log(x_1^2 + x_2^2)$ as a distribution in $\mathbb{R}^2$ does not satisfy the Laplace equation. In fact

$$\frac{\partial^2 u}{\partial x_1^2} + \frac{\partial^2 u}{\partial x_2^2} = c\delta$$

(and in $\mathbb{R}^3$ similarly

$$\frac{\partial^2 v}{\partial x_1^2} + \frac{\partial^2 v}{\partial x_2^2} + \frac{\partial^2 v}{\partial x_3^2} = c_1 \delta$$

for $v(x_1, x_2, x_3) = (x_1^2 + x_2^2 + x_3^2)^{-1/2})$ for certain constants c and c_1. These facts are not just curiosities—they are useful in solving Poisson's equation

$$\frac{\partial^2 u}{\partial x_1^2} + \frac{\partial^2 u}{\partial x_2^2} = f,$$

as we shall see.

1.4   Problems

1. Let H be the distribution in $\mathcal{D}'(\mathbb{R}^1)$ defined by the Heaviside function

$$H(x) = \begin{cases} 1 & \text{if } x > 0 \\ 0 & \text{if } x \leq 0. \end{cases}$$

   Show that if $h_n(x)$ are differentiable functions such that $\int h_n(x)\varphi(x)\,dx \to \langle H, \varphi \rangle$ as $n \to \infty$ for all $\varphi \in \mathcal{D}$, then

$$\int h'_n(x)\varphi(x)\,dx \to \langle \delta, \varphi \rangle.$$

   Does the answer change if you redefine $H(x)$ to be 1 at $x = 0$?

2. Let f_n be the distribution $\langle f_n, \varphi \rangle = n\left(\varphi\left(\frac{1}{n}\right) - \varphi\left(-\frac{1}{n}\right)\right)$. What distribution is $\lim_{n\to\infty}\langle f_n, \varphi \rangle$?

3. For any $a > 0$ show that

$$\langle f_a, \varphi \rangle = \int_{-\infty}^{-a} + \int_{a}^{\infty} \frac{\varphi(x)}{|x|}\,dx + \int_{-a}^{a} \frac{\varphi(x) - \varphi(0)}{|x|}\,dx$$

   is a distribution. (*Hint*: the problem is the convergence of the last integral near $x = 0$. Use the mean value theorem.)

4. Show $\langle f_a, \varphi \rangle = \int_{-\infty}^{\infty} \frac{\varphi(x)}{|x|}\,dx$ for any $\varphi \in \mathcal{D}(\mathbb{R}^1)$ for which $\varphi(0) = 0$.

5. What distribution is $f_a - f_b$ for $a > b > 0$?

6. For any $a > 0$ show

$$\langle g_a, \varphi \rangle = \int_{-\infty}^{-a} + \int_{a}^{\infty} \frac{\varphi(x)}{x}\,dx + \int_{-a}^{a} \frac{\varphi(x) - \varphi(0)}{x}\,dx$$

   is a distribution.

7. Prove that g_a does not depend on a, and the resulting distribution may also be given by

$$\lim_{a\to 0^+} \int_{-\infty}^{-a} + \int_{a}^{\infty} \frac{\varphi(x)}{x}\,dx.$$

8. Suppose f is a distribution on $\mathbb{R}^1$. Show that $\langle F, \varphi \rangle = \langle f, \varphi_y \rangle$, for $\varphi \in \mathcal{D}(\mathbb{R}^2)$, where $\varphi_y(x) = \varphi(x, y)$ (here y is any fixed value), defines a distribution on $\mathbb{R}^2$.

9. Suppose f is a distribution on $\mathbb{R}^1$. Show that $\langle G, \varphi \rangle = \int_{-\infty}^{\infty} \langle f, \varphi_y \rangle \, dy$ for $\varphi \in \mathcal{D}(\mathbb{R}^2)$ defines a distribution on $\mathbb{R}^2$. Is G the same as F in problem *8*?

10. Show that

$$\langle f, \varphi \rangle = \sum_{n=1}^{\infty} \varphi^{(n)}(n)$$

   defines a distribution on $\mathbb{R}^1$. (*Hint:* Is it really an infinite series?)

11. Why doesn't $\langle f, \varphi \rangle = \varphi(0)^2$ define a distribution?

12. Show that the line integral $\int_C \varphi \, ds$ along a smooth curve C in the plane defines a distribution on $\mathbb{R}^2$. Do the same for surface integrals in $\mathbb{R}^3$.

13. Working with complex-valued test functions and distributions, define f to be real-valued if $\overline{\langle f, \varphi \rangle} = \langle f, \bar{\varphi} \rangle$. Show that $\langle \operatorname{Re} f, \varphi \rangle = \frac{1}{2}(\langle f, \varphi \rangle + \overline{\langle f, \bar{\varphi} \rangle})$ and $\langle \operatorname{Im} f, \varphi \rangle = \frac{1}{2i}(\langle f, \varphi \rangle - \overline{\langle f, \bar{\varphi} \rangle})$ define real-valued distributions, and $f = \operatorname{Re} f + i \operatorname{Im} f$.

14. Suppose f and g are distributions such that $\langle f, \varphi \rangle = 0$ if and only if $\langle g, \varphi \rangle = 0$. Show that $\langle f, \varphi \rangle = c \langle g, \varphi \rangle$ for some constant c.

Chapter 2

The Calculus of Distributions

2.1 Functions as distributions

The next thing you have to learn is how to calculate with distributions. By now you should have the idea that distributions are some sort of "function-like" objects, some actually being functions and some like δ and δ', definitely not functions. Things that you are used to doing with functions, such as adding them or differentiating them, should be possible with distributions. You should also expect to get the same answer if you regard a function as a function or a distribution.

Now let me backtrack a little and discuss more carefully the "identification" of some functions with some distributions. If f is a function such that the integral $\int f(x)\varphi(x)\,dx$ exists for every test function (the integral must exist in an absolutely convergent sense—$\int |f(x)\varphi(x)|\,dx$ must be finite—although it may be an improper integral because f becomes infinite at some points), then

$$\langle f, \varphi \rangle = \int f(x)\varphi(x)\,dx$$

defines a distribution. Do two different functions define the same distribution? They may if they are equal except on a set that is so small it does not affect the integral (in terms of Lebesgue integration theory, if they are equal almost everywhere). For instance, if

$$f(x) = \begin{cases} 0 & \text{if } x \neq 0 \\ 1 & \text{if } x = 0 \end{cases}$$

12

then $\int f(x)\varphi(x)\,dx = 0$ for all test function, so f and the zero function define the same distribution (call it the zero distribution). Intuitively this makes sense—a function that vanishes everywhere except at the origin must also vanish at the origin. Anything else is an experimental error. Already you see how distribution theory forces you to overlook useless distinctions! But if the functions f_1 and f_2 are "really" different, say $f_1 > f_2$ on some interval, then the distributions f_1 and f_2 are really different: $\langle f_1, \varphi \rangle > \langle f_2, \varphi \rangle$ if φ is a nonnegative test function vanishing outside the interval.

Now for some notation. A function $f(x)$ defined on Ω for which $\int f(x)\varphi(x)\,dx$ is absolutely convergent for every $\varphi \in \mathcal{D}(\Omega)$ is called *locally integrable*, denoted $f \in L^1_{\text{loc}}(\Omega)$. A criterion for local integrability is the finiteness of the integral $\int_B |f(x)|\,dx$ over all sets $B \subseteq \Omega$ that are bounded and stay away from the boundary of Ω. To explain this terminology let me mention that an *integrable* function on $\Omega(f \in L^1(\Omega))$ is one for which $\int_\Omega |f(x)|\,dx$ is finite. Clearly an integrable function is locally integrable but not conversely. For example $f(x) \equiv 1$ is locally integrable but not integrable on $\mathbb{R}^n$, and $f(x) = \frac{1}{x}$ is locally integrable but not integrable on $0 < x < \infty$.

For every locally integrable function f we associate a distribution, also denoted f, given by $\langle f, \varphi \rangle = \int f(x)\varphi(x)\,dx$. We say that two locally integrable functions are *equivalent* if as distributions they are equal. Thus by ignoring the distinction between equivalent functions, we can regard the locally integrable functions as a subset of the distributions, $L^1_{\text{loc}}(\Omega) \subseteq \mathcal{D}'(\Omega)$. This makes precise the intuitive statement that the distributions are a set of objects larger than the set of functions, justifying the term "generalized functions."

There are many interesting functions that are not locally integrable, for instance $1/x$ and $1/|x|$ on $\mathbb{R}^1$ (you should not be upset that I said recently that $1/x$ is locally integrable on $0 < x < \infty$; the concept of local integrability depends on the set Ω being considered—Ω determines what "local" means). In the problems you have already encountered distributions associated with these functions; for example, a distribution satisfying

$$\langle f, \varphi \rangle = \int \frac{\varphi(x)}{|x|}\,dx,$$

whenever $\varphi(0) = 0$. However, *this is an entirely different construction. You should never expect that properties or operations concerning the function carry over to the distribution, unless the function is locally integrable.* For instance, you have seen that more than one distribution corresponds to $1/|x|$—in fact, it is impossible to single out one as more "natural" than another. Also, although the function $1/|x|$ is nonnegative, none of the associated distributions are nonnegative (a distribution f is nonnegative if

$\langle f, \varphi \rangle \geq 0$ for every test function φ that satisfies $\varphi(x) \geq 0$ for all $x \in \Omega$). In contrast, a nonnegative locally integrable function *is* nonnegative as a distribution. (Exercise: Verify this.) Thus you see that *this is an aspect of distribution theory where it is very easy to make mistakes.*

2.2   Operations on distributions

Now we are ready to move forward. We have $L^1_{\text{loc}}(\Omega) \subseteq \mathcal{D}'(\Omega)$ and hence $\mathcal{D}(\Omega) \subseteq \mathcal{D}'(\Omega)$ since every test function is integrable. Any property or operation on test functions may be extended to distributions. Now why did I say "test functions" rather then "locally integrable functions"? Simply because there are more things you can do with them: for example, differentiation. How does the extension to distributions work? There are essentially two ways to proceed. Both always give the same result.

At this point I will make the convention that *operation* means linear operator (or linear transform), that is, for any $\varphi \in \mathcal{D}, T\varphi \in \mathcal{D}$ and

$$T(a_1\varphi_1 + a_2\varphi_2) = a_1 T\varphi_1 + a_2 T\varphi_2.$$

At times we will want to consider operations $T\varphi$ that may not yield functions in $\mathcal{D}$, but we will always want linearity to hold.

The first way to proceed is to approximate an arbitrary distribution by test functions. That this is always possible is a remarkable and basic fact of the theory. We say that a sequence of distributions $f_1, f_2, \ldots$ converges to the distribution f if the sequence $\langle f_n, \varphi \rangle$ converges to the number $\langle f, \varphi \rangle$ for all test functions. Write this $f_n \to f$, or say simply that the f_n approximate f. Incidentally, you can use the limiting process to construct distributions. If $\{f_n\}$ is a sequence of distributions for which the limit as $n \to \infty$ of $\langle f_n, \varphi \rangle$ exists for all test functions φ, then $\langle f, \varphi \rangle = \lim_{n \to \infty} \langle f_n, \varphi \rangle$ defines a distribution and of course $f_n \to f$.

Theorem 2.2.1 *Given any distribution $f \in \mathcal{D}'(\Omega)$, there exists a sequence $\{\varphi_n\}$ of test functions such that $\varphi_n \to f$ as distributions.*

Accepting this as true, if T is any operation defined on test functions we may extend T to distributions as follows:

Meta-Definition 1: $Tf = \lim_{n \to \infty} T\varphi_n$ if $\{\varphi_n\}$ is a sequence of test functions such that $f = \lim_{n \to \infty} \varphi_n$.

Of course for this to make sense, $\lim_{n \to \infty} T\varphi_n$ must be independent of the choice of the sequence $\{\varphi_n\}$ approximating f. Fortunately, this is the case for most interesting operators—it is a consequence of continuity.

To see how this definition works, let us consider two examples. In the first, let T be a translation $T\varphi(x) = \varphi(x + y)$, y a fixed vector in $\mathbb{R}^n$, and $\varphi \in \mathcal{D}(\mathbb{R}^n)$. Now let $f \in \mathcal{D}(\mathbb{R}^n)$ be a distribution. How shall we define Tf? We must say what $\langle Tf, \varphi \rangle$ is for an arbitrary test function φ. So we approximate $f = \lim_{n \to \infty} \varphi_n$ for $\varphi_n \in \mathcal{D}(\mathbb{R}^n)$, which means

$$\langle f, \varphi \rangle = \lim_{n \to \infty} \int \varphi_n(x) \varphi(x) \, dx.$$

According to the meta-definition, $Tf = \lim_{n \to \infty} T\varphi_n$, so

$$(Tf, \varphi) = \lim_{n \to \infty} \int T\varphi_n(x) \varphi(x) \, dx.$$

Since $T\varphi_n(x) = \varphi_n(x + y)$ we can write this as

$$\lim_{n \to \infty} \int \varphi_n(x + y) \varphi(x) \, dx.$$

If f is the Dirac δ-function $f = \delta, \langle \delta, \varphi \rangle = \varphi(0)$, we may take the sequence φ_n to look like

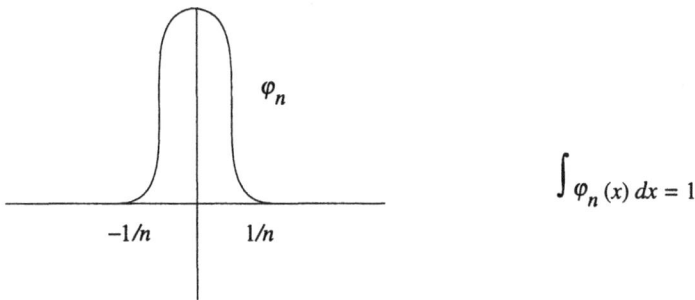

Figure 2.1

Then $\varphi_n(x + y)$ as a function of x is concentrated near $x = -y$ so that

$$\lim_{n \to \infty} \int \varphi_n(x + y) \varphi(x) \, dx = \varphi(-y).$$

Thus $\langle T\delta, \varphi \rangle = \varphi(-y)$. (This is sometimes called the δ-function at $-y$.)

In general we can do a similar simplification by making a change of variable $x \to x - y$:

$$\langle Tf, \varphi \rangle = \lim_{n \to \infty} \int \varphi_n(x+y)\varphi(x)\,dx$$

$$= \lim_{n \to \infty} \int \varphi_n(x)\varphi(x-y)\,dx$$

$$= \langle f, \varphi(x-y) \rangle.$$

Paraphrased: To translate a distribution through y, translate the test function through $-y$.

If you are puzzled by the appearance of the minus sign, please go over the above manipulations again. It will show up in later computations as well (also look back at the definition of δ').

Next example: $T = d/dx$ (for simplicity assume $n = 1$). This is the infinitesimal version of translation. If we write $\tau_y\varphi(x) = \varphi(x+y)$ for translation and $I\varphi(x) = \varphi(x)$ for the identity, then

$$\frac{d}{dx} = \lim_{y \to 0} \frac{1}{y}(\tau_y - I).$$

So we expect to have

$$\frac{d}{dx}f = \lim_{y \to 0}\frac{1}{y}(\tau_y f - f).$$

Since

$$\langle \tau_y f, \varphi \rangle = \langle f, \tau_{-y}\varphi \rangle$$

(this was the first example) we have

$$\left\langle \frac{d}{dx}f, \varphi \right\rangle = \lim_{y \to 0}\frac{1}{y}(\langle f, \tau_{-y}\varphi \rangle - \langle f, \varphi \rangle)$$

$$= \lim_{y \to 0}\left\langle f, \frac{1}{y}(\tau_{-y}\varphi - \varphi) \right\rangle.$$

But

$$\lim_{y \to 0}\frac{1}{y}(\tau_{-y}\varphi - \varphi) = \lim_{y \to 0}\frac{1}{y}(\varphi(x-y) - \varphi(x)) = -\varphi'(x)$$

(there's that minus sign again). So it stands to reason that

$$\left\langle \frac{d}{dx}f, \varphi \right\rangle = -\left\langle f, \frac{d}{dx}\varphi \right\rangle$$

(notice that this defines a distribution since $(d/dx)\varphi$ is a test function whenever φ is).

Let's do that again more directly. Let $\varphi_n \to f, \varphi_n \in \mathcal{D}(\mathbb{R}^1)$. Then $\frac{d}{dx}\varphi_n \to \frac{d}{dx}f$, meaning

$$\left\langle \frac{d}{dx}f, \varphi \right\rangle = \lim_{n\to\infty} \int \left(\frac{d}{dx}\varphi_n(x) \right) \varphi(x)\,dx.$$

To simplify this, we integrate by parts (this is analogous to the change of variable $x \to x - y$ in the first example):

$$\int \left(\frac{d}{dx}\varphi_n(x) \right) \varphi(x)\,dx = - \int \varphi_n(x)\frac{d}{dx}\varphi(x)\,dx$$

(there are no boundary terms because $\varphi(x) = 0$ outside a bounded set). Substituting in we obtain

$$\begin{aligned}
\left\langle \frac{d}{dx}f, \varphi \right\rangle &= \lim_{n\to\infty} \int \varphi_n(x)\frac{d}{dx}\varphi(x)\,dx \\
&= -\left\langle f, \frac{d}{dx}\varphi \right\rangle.
\end{aligned}$$

Again it is instructive to look at the special case $f = \delta$. Then we have $\varphi_n \to \delta$ where

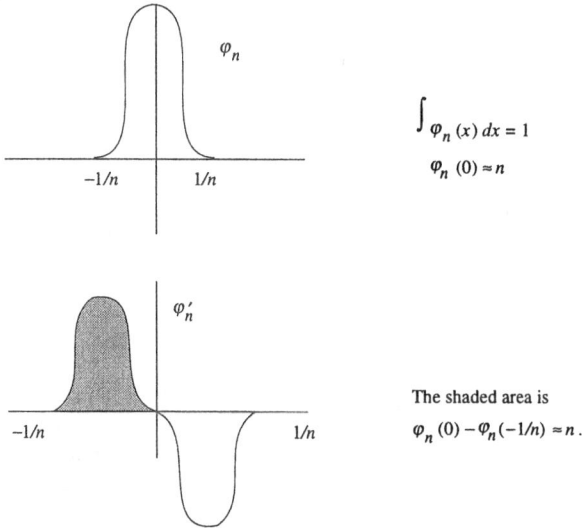

$$\int \varphi_n(x)\,dx = 1$$
$$\varphi_n(0) \approx n$$

The shaded area is
$$\varphi_n(0) - \varphi_n(-1/n) \approx n.$$

Figure 2.2

So $\int \varphi_n(x)\varphi(x)\,dx \approx n\left(\varphi\left(\frac{-1}{2n}\right)\right) - \varphi\left(\left(\frac{1}{2n}\right)\right) \to -\varphi'(0)$ as $n \to \infty$.

Exercise: Verify that for $f \in \mathcal{D}'(\Omega), \Omega \subseteq \mathbb{R}^n$, we have

$$\left\langle \frac{\partial}{\partial x_k} f, \varphi \right\rangle = - \left\langle f, \frac{\partial}{\partial x_k} \varphi \right\rangle.$$

2.3 Adjoint identities

Now let us turn to the second method of defining operations on distribution, the method of *adjoint identities*. An adjoint identity looks like[1]

$$\int T\psi(x)\varphi(x)\,dx = \int \psi(x)S\varphi(x)\,dx$$

for all $\varphi, \psi \in \mathcal{D}$. Here T is the operator we are interested in, and S is an operator that makes the identity true (in other words, there is in general no recipe for S in terms of T). Suppose we are lucky enough to discover an adjoint identity involving T. Then we may simply define $\langle Tf, \varphi \rangle = \langle f, S\varphi \rangle$. The adjoint identity says this is true if $f \in \mathcal{D}$, and since any distribution f is the limit of test functions, $\varphi_n \to f, \varphi_n \in \mathcal{D}$, and we have $\langle T\varphi_n, \varphi \rangle = \langle \varphi_n, S\varphi \rangle$ by the adjoint identity, we obtain

$$\lim_{n\to\infty} \langle T\varphi_n, \varphi \rangle = \lim_{n\to\infty} \langle \varphi_n, S\varphi \rangle$$
$$= \langle f, S\varphi \rangle.$$

Thus, *if an adjoint identity exists*, then we must have

$$\langle Tf, \varphi \rangle = \langle f, S\varphi \rangle$$

by the first definition. This also shows that the answer does not depend on the choice of the approximating sequence, something that was not obvious before.

In most cases we will use this second definition and completely by-pass the first definition. But that means we first have to do the work of finding the adjoint identity. Another point we must bear in mind is that *we must have $S\varphi \in \mathcal{D}$ whenever $\varphi \in \mathcal{D}$*, otherwise $\langle f, S\varphi \rangle$ may not be defined.

Now let us look at the two previous examples from this point of view. In both cases the final answer gives us the adjoint identity if we specialize to $f \in \mathcal{D}$:

$$\int \tau_y\psi(x)\varphi(x)\,dx = \int \psi(x)\tau_{-y}\varphi(x)\,dx$$

---

[1]Note that in Hilbert space theory the adjoint identity involves complex conjugates, while in distribution theory we do not take complex conjugates even when we deal with complex valued functions.

and

$$\int \left(\frac{d}{dx}\psi(x)\right)\varphi(x)\,dx = -\int \psi(x)\frac{d}{dx}\varphi(x)\,dx.$$

Of course both of these are easy to verify directly, the first by the change of variable $x \to x - y$ and the second by integration by parts. From these we get back

$$\langle \tau_y f, \varphi \rangle = \langle f, \tau_{-y}\varphi \rangle$$

and

$$\left\langle \frac{d}{dx}f, \varphi \right\rangle = -\left\langle f, \frac{d}{dx}\varphi \right\rangle.$$

One more example: multiplication by a function $m(x)$. The adjoint identity is obvious:

$$\int (m(x)\psi(x))\varphi(x)\,dx = \int \psi(x)(m(x)\varphi(x))\,dx.$$

However, here we must be careful that $m(x)\varphi(x) \in \mathcal{D}$ whenever $\varphi(x) \in \mathcal{D}$. This is not always the case—what we need to assume is that $m(x)$ is infinitely differentiable (it is not necessary to assume anything about the "size" of $m(x)$ at ∞ or near the boundary of Ω). In that case we have

$$\langle m \cdot f, \varphi \rangle = \langle f, m \cdot \varphi \rangle.$$

Note that if m is discontinuous at $x = 0$, say $m(x) = \operatorname{sgn} x$, then $m \cdot \delta$ cannot be sensibly defined, for by rights we should have $\langle m \cdot \delta, \varphi \rangle = \varphi(0)m(0)$, but there is no consistent way to define $m(0)$. Thus *the product of two arbitrary distributions is undefined.*

Another thing to keep in mind is that the product $m \cdot f$ for $m \in C^\infty$ may not be what you first think it is. For instance, let's compute $m \cdot \delta'$. We have

$$\begin{aligned}
\langle m \cdot \delta', \varphi \rangle &= \langle \delta', m \cdot \varphi \rangle \\
&= -\left\langle \delta, \frac{d}{dx}(m\varphi) \right\rangle \\
&= -\langle \delta, m'\varphi \rangle - \langle \delta, m\varphi' \rangle \\
&= -m'(0)\varphi(0) - m(0)\varphi'(0) \\
&= \langle -m'(0)\delta + m(0)\delta', \varphi \rangle
\end{aligned}$$

so $m \cdot \delta' = m(0)\delta' - m'(0)\delta$ (not just $m(0)\delta'$ as you may have guessed).

Now let us pause to consider the momentous consequences of what we have done. We start with the class of locally integrable functions. We enlarge it to the class of distribution. Suddenly everything has derivatives!

Of course if $f \in L^1_{\text{loc}}$ is regarded as a distribution, then the distribution $(\partial/\partial x_k)f$ need not correspond to a locally integrable function. Nevertheless, something called $(\partial/\partial x_k)f$ exists, and we can perform operations on it and postpone the question of "what exactly is it" until all computations are completed. This is the point of view that has revolutionized the theory of differential equations. In this sense we are justified in claiming that distribution theory is the completion of differential calculus.

2.4   Consistency of derivatives

At this point a very natural question arises: if $(\partial/\partial x_k)f$ exists in the ordinary sense, does it agree with $(\partial/\partial x_k)f$ in the distribution sense? Fortunately the answer is yes, as long as $(\partial/\partial x_k)f$ exists as a *continuous* derivative.

To see this we had better change notation temporarily. Suppose $f(x) \in L^1_{\text{loc}}$, and let

$$g(x) = \lim_{h \to 0} \frac{1}{h}(f(x + he_k) - f(x))$$

(e_k is the unit vector in the x_k-direction) exist and be continuous. Then

$$\int g(x)\varphi(x)\, dx = -\int f(x)\frac{\partial}{\partial x_k}\varphi(x)\, dx$$

for any test function by the integration-by-parts formula in the x_k-integral. However, the distribution $(\partial/\partial x_k)f$ is defined by

$$\left\langle \frac{\partial}{\partial x_k}f, \varphi \right\rangle = -\left\langle f, \frac{\partial}{\partial x_k}\varphi \right\rangle$$

$$= -\int f(x)\frac{\partial}{\partial x_k}\varphi(x)\, dx$$

so $(\partial/\partial x_k)f = g$ as distributions. Thus the two notions coincide.

You can go a little farther, as the following theorem asserts:

Theorem 2.4.1 *Let $f(x) \in L^1_{\text{loc}}(\mathbb{R}^n)$ and let $g(x)$ as above exist and be continuous except as a single point y, and let $g \in L^1_{\text{loc}}(\mathbb{R}^n)$ (define it arbitrarily at the point y). Then, if $n \geq 2$ the distribution derivative $\partial f/\partial x_k$ equals g, while if $n = 1$ this is true if, in addition, f is continuous at y.*

The hypothesis of continuity when $n = 1$ is needed because $d/dx \operatorname{sgn} x = 0$ for all $x \neq 0$ in the ordinary sense but $d/dx \operatorname{sgn} x = 2\delta$ in the distribution sense.

Let me indicate a brief proof. For simplicity, take $y = 0$. First for $n = 1$ we have

$$\left\langle \frac{d}{dx} f, \varphi \right\rangle = -\left\langle f, \frac{d}{dx}\varphi \right\rangle$$

$$= -\int_{-\infty}^{\infty} f(x)\varphi'(x)\,dx$$

$$= -\lim_{\epsilon \to 0}\int_{-\infty}^{-\epsilon} + \int_{\epsilon}^{\infty} f(x)\varphi'(x)\,dx.$$

We have cut away a neighborhood of the exceptional point. Now we can apply integration by parts:

$$\int_{-\infty}^{-\epsilon} f(x)\varphi'(x)\,dx = f(-\epsilon)\varphi(-\epsilon) - \int_{-\infty}^{-\epsilon} g(x)\varphi(x)\,dx$$

and

$$\int_{\epsilon}^{\infty} f(x)\varphi'(x)\,dx = -f(\epsilon)\varphi(\epsilon) - \int_{\epsilon}^{\infty} g(x)\varphi(x)\,dx.$$

This time there are boundary terms, but they appear with opposite sign. Adding, we obtain

$$f(-\epsilon)\varphi(-\epsilon) - f(\epsilon)\varphi(\epsilon) - \left(\int_{-\infty}^{-\epsilon} + \int_{\epsilon}^{\infty} g(x)\varphi(x)\,dx\right).$$

Now let $\epsilon \to 0$. The first two terms approach $f(0)\varphi(0) - f(0)\varphi(0) = 0$ because f is continuous. Thus

$$\left\langle \frac{d}{dx} f, \varphi \right\rangle = \lim_{\epsilon \to 0}\int_{-\infty}^{-\epsilon} + \int_{\epsilon}^{\infty} g(x)\varphi(x)\,dx$$

$$= \int_{-\infty}^{\infty} g(x)\varphi(x)\,dx$$

because g is locally integrable.

Exercise: What happens if f has a jump discontinuity at 0?

For $n \geq 2$ the argument is simpler. Let's do $n = 2$ and $\partial/\partial x_1$ for instance. We have

$$\left\langle \frac{\partial}{\partial x_1} f, \varphi \right\rangle = -\int f(x)\frac{\partial \varphi}{\partial x_1}(x)\,dx.$$

We write the double integral as an iterated integral, doing the x_1-integration first:

$$-\int f(x)\frac{\partial \varphi}{\partial x_1}(x)\,dx = -\int_{-\infty}^{\infty}\left(\int_{-\infty}^{\infty} f(x_1, x_2)\frac{\partial \varphi}{\partial x_1}(x_1, x_2)\,dx_1\right)dx_2.$$

Now for every x_2, except $x_2 = 0$, $f(x_1, x_2)$ is continuously differentiable in x_1, so we integrate by parts:

$$\int_{-\infty}^{\infty} f(x_1, x_2) \frac{\partial \varphi}{\partial x_1}(x_1, x_2)\, dx_1 = -\int_{-\infty}^{\infty} g(x_1, x_2) \varphi(x_1, x_2)\, dx_1.$$

Putting this back in the iterated integral gives what we want since the single point $x_2 = 0$ does not contribute to the integral.

2.5 Distributional solutions of differential equations

Now we are in a position to discuss solutions to differential equations from a distribution point of view. Remember the vibrating strings equation:

$$\frac{\partial^2 u(x, t)}{\partial t^2} = k^2 \frac{\partial^2 u(x, t)}{\partial x^2} \quad ?$$

Both sides of the equation are meaningful if u is a distribution. If the equality holds, we call u a *weak solution*. If u is twice continuously differentiable and the equality holds, we call u a *classical solution*. By what we have just seen, a classical solution is also a weak solution. From a physical point of view there is no reason to prefer classical solutions to weak solutions.

Now consider a traveling wave $u(x, t) = f(x - kt)$ for $f \in L^1_{\text{loc}}(\mathbb{R}^2)$. It is easy to see that $u \in L^1_{\text{loc}}(\mathbb{R}^2)$ so it defines a distribution. Is it a weak solution?

$$\left\langle \frac{\partial^2}{\partial t^2} u, \varphi \right\rangle = \left\langle u, \frac{\partial^2}{\partial t^2}\varphi \right\rangle = \iint f(x - kt) \frac{\partial^2}{\partial t^2}\varphi(x, t)\, dx\, dt$$

and

$$\left\langle \frac{\partial^2}{\partial x^2} u, \varphi \right\rangle = \left\langle u, \frac{\partial^2}{\partial x^2}\varphi \right\rangle = \iint f(x - kt) \frac{\partial^2}{\partial x^2}\varphi(x, t)\, dx\, dt$$

so

$$\left\langle \frac{\partial^2}{\partial t^2} u - k^2 \frac{\partial^2}{\partial x^2} u, \varphi \right\rangle = \iint f(x - kt) \left(\frac{\partial^2}{\partial t^2}\varphi(x, t) - k^2 \frac{\partial^2}{\partial x^2}\varphi(x, t) \right) dx\, dt.$$

We want this to be zero. To see this we introduce new variables $y = x - kt$ and $z = x + kt$, so

$$\frac{\partial(y, z)}{\partial(x, t)} = \begin{pmatrix} 1 & -k \\ 1 & k \end{pmatrix} \qquad dx\, dt = \frac{1}{2k} dy\, dz.$$

Then

$$\frac{\partial}{\partial x} = \frac{\partial}{\partial y} + \frac{\partial}{\partial z} \quad \text{and} \quad \frac{\partial}{\partial t} = -k\frac{\partial}{\partial y} + k\frac{\partial}{\partial z}$$

so

$$\frac{\partial^2}{\partial t^2} - k^2\frac{\partial^2}{\partial x^2} = -4k^2\frac{\partial}{\partial y}\frac{\partial}{\partial z}.$$

Thus

$$\iint f(x - kt)\left(\frac{\partial^2}{\partial t^2} - k^2\frac{\partial^2}{\partial x^2}\right)\varphi(x,t)\,dx\,dt$$

$$= -2k\iint f(y)\frac{\partial^2\varphi}{\partial y\partial z}(y,z)\,dz\,dy.$$

We claim the z integration already produces zero, for all y. To see this note that

$$\int_a^b \frac{\partial^2\varphi}{\partial y\partial z}(y,z)\,dz = \frac{\partial\varphi}{\partial y}(y,b) - \frac{\partial\varphi}{\partial y}(y,a).$$

Thus

$$\int_{-\infty}^\infty \frac{\partial^2\varphi}{\partial y\partial z}(y,z)\,dz = 0$$

since φ and hence $\partial\varphi/\partial y$ vanishes outside a bounded set. Thus $u(x,t) = f(x - kt)$ is a weak solution.

In contrast, let's see if $\log(x^2 + y^2)$ is a weak solution to the Laplace equation

$$\left(\frac{\partial^2}{\partial x^2} + \frac{\partial^2}{\partial y^2}\right)u(x,y) = 0.$$

For u to be a weak solution means

$$\left\langle \left(\frac{\partial^2}{\partial x^2} + \frac{\partial^2}{\partial y^2}\right)u(x,y), \varphi(x,y)\right\rangle$$

$$\left\langle u, \left(\frac{\partial^2}{\partial x^2} + \frac{\partial^2}{\partial y^2}\right)\varphi\right\rangle = 0$$

for all test functions φ. It will simplify matters if we pass to polar coordinates, in which case

$$\frac{\partial^2}{\partial x^2} + \frac{\partial^2}{\partial y^2} = \frac{\partial^2}{\partial r^2} + \frac{1}{r}\frac{\partial}{\partial r} + \frac{1}{r^2}\frac{\partial^2}{\partial\theta^2}$$

and $dx\,dy = r\,dr\,d\theta$. Then for $u = \log x^2 + y^2 = \log r^2$ the question boils down to

$$\int_0^{2\pi}\int_0^\infty \log r^2\left(\left(\frac{\partial^2}{\partial r^2} + \frac{1}{r}\frac{\partial}{\partial r} + \frac{1}{r^2}\frac{\partial^2}{\partial\theta^2}\right)\varphi(r,\theta)\right)r\,dr\,d\theta = 0?$$

To avoid the singularity of u at the origin we extend the r-integration from ϵ to ∞ for $\epsilon > 0$ and let $\epsilon \to 0$. Before letting $\epsilon \to 0$ we integrate by parts to put all the derivatives on u. Since u is harmonic away from the origin we expect to get zero from the integral term, but the boundary terms will not be zero and they will determine what happens. So let's compute.

$$\int_0^{2\pi} \log r^2 \frac{1}{r^2} \frac{\partial^2}{\partial\theta^2}\varphi(r,\theta) r\, d\theta = \frac{1}{r}\log r^2 \frac{\partial\varphi}{\partial\theta}(r,\theta)\Big|_0^{2\pi} = 0$$

since $\partial\varphi/\partial\theta$ is periodic. So that term is always zero. Next

$$\int_\epsilon^\infty \log r^2 \frac{\partial}{\partial r}\varphi(r,\theta)\, dr = -\int_\epsilon^\infty \left(\frac{\partial}{\partial r}\log r^2\right)\varphi(r,\theta)\, dr - \log\epsilon^2\varphi(\epsilon,\theta).$$

Finally

$$\int_\epsilon^\infty r\log r^2 \frac{\partial^2}{\partial r^2}\varphi(r,\theta)\, dr = -\int_\epsilon^\infty \left(\frac{\partial}{\partial r}(r\log r^2)\right)\frac{\partial}{\partial r}\varphi(r,\theta)\, dr$$

$$-\epsilon\log\epsilon^2 \frac{\partial}{\partial r}\varphi(\epsilon,\theta)$$

$$= \int_\epsilon^\infty \left(\frac{\partial^2}{\partial r^2}(r\log r^2)\right)\varphi(r,\theta)\, dr - \epsilon\log\epsilon^2 \frac{\partial\varphi}{\partial r}(\epsilon,\theta)$$

$$+ \frac{\partial}{\partial r}(r\log r^2)\Big|_{r=\epsilon}\varphi(\epsilon,\theta).$$

Now

$$\frac{\partial}{\partial r}(r\log r^2) = \log r^2 + 2, \quad \frac{\partial^2}{\partial r^2}(r\log r^2) = \frac{2}{r} \quad \text{and} \quad \frac{\partial}{\partial r}\log r^2 = \frac{2}{r}$$

so adding everything up we obtain

$$\int_0^{2\pi}\int_\epsilon^\infty \log r^2 \left(\frac{\partial^2}{\partial r^2} + \frac{1}{r}\frac{\partial}{\partial r} + \frac{1}{r^2}\frac{\partial^2}{\partial\theta^2}\right)\varphi(r,\theta) r\, dr\, d\theta$$

$$= \int_0^{2\pi}\int_\epsilon^\infty \left(-\frac{2}{r} + \frac{2}{r}\right)\varphi(r,\theta)\, dr\, d\theta$$

$$+ \int_0^{2\pi}(-\log\epsilon^2 + \log\epsilon^2 + 2)\varphi(\epsilon,\theta)\, d\theta$$

$$+ \int_0^{2\pi}(-\epsilon\log\epsilon^2)\frac{\partial\varphi}{\partial r}(\epsilon,\theta)\, d\theta.$$

The first term is zero, as we expected, so

$$\langle\Delta u,\varphi\rangle = \lim_{\epsilon\to 0} 2\int_0^{2\pi}\varphi(\epsilon,\theta)\, d\theta - \int_0^{2\pi}\epsilon\log\epsilon^2 \frac{\partial\varphi}{\partial r}(\epsilon,\theta)\, d\theta.$$

Since φ is continuous, $\varphi(\epsilon,\theta)$ approaches the value of φ at the origin as $\epsilon \to 0$. Thus the first term approaches $4\pi\langle\delta,\varphi\rangle$. In the second term $\partial\varphi/\partial r$ remains bounded while $\epsilon\log\epsilon^2 \to 0$ as $\epsilon \to 0$ so the limit is zero. Thus $\Delta\log(x^2+y^2) = 4\pi\delta$, so $\log(x^2+y^2)$ is not a weak solution of $\Delta u = 0$.

Actually the above computation is even more significant, for it will enable us to solve the equation $\Delta u = f$ for any function f. We will return to this later.

2.6 Problems

1. Extend to distributions the dilations $d_r\varphi(x) = \varphi(rx_1, rx_2, \ldots, rx_n)$ for $r > 0$. (*Hint* : $\int (d_r\psi(x))\varphi(x)\,dx = r^{-n}\int \psi(x)d_{r^{-1}}\varphi(x)\,dx$.)) Show that k $f(x) = |x|^t$ for $t > -n$ is homogeneousof degree t (meaning $d_r f = r^t f$). Show that the distribution

$$\langle f,\varphi\rangle = \int_{-\infty}^{-1} + \int_1^\infty \frac{\varphi(x)}{x}\,dx$$
$$+ \int_{-1}^1 \frac{\varphi(x)-\varphi(0)}{x}\,dx$$

is homogeneous of degree -1, but

$$\langle g,\varphi\rangle = \int_{-\infty}^{-1} + \int_1^\infty \frac{\varphi(x)}{|x|}\,dx + \int_{-1}^1 \frac{\varphi(x)-\varphi(0)}{|x|}\,dx$$

is not homogeneous of degree -1.

2. Prove that δ is homogeneous of degree $-n$.

3. Prove that
$$\frac{\partial}{\partial x_k}\frac{\partial}{\partial x_j}f = \frac{\partial}{\partial x_j}\frac{\partial}{\partial x_k}f$$
for any distribution f.

4. Prove
$$\frac{\partial}{\partial x_j}(m\cdot f) = \frac{\partial m}{\partial x_j}\cdot f + m\frac{\partial}{\partial x_j}f$$
for any distribution f and C^∞ function m.

5. Show that if f is homogeneous of degree t then $x_j\cdot f$ is homogeneous of degree $t+1$ and $(\partial/\partial x_j)f$ is homogeneous of degree $t-1$.

6. Define $\tilde{\varphi}(x) = \varphi(-x)$ and $\langle \tilde{f}, \varphi \rangle = \langle f, \tilde{\varphi} \rangle$. Show this definition is consistent for distributions defined by functions. What is $\tilde{\delta}$? Call a distribution even if $\tilde{f} = f$ and odd if $\tilde{f} = -f$. Show that every distribution is uniquely the sum of an even and odd distribution (*Hint:* Find a formula for the even and odd part.) Show that $\langle f, \varphi \rangle = 0$ if f and φ have opposite parity (i.e., one is odd and the other is even).

7. Let h be a C^∞ function from $\mathbb{R}^1$ to $\mathbb{R}^1$ with $h'(x) \neq 0$ for all x. How would you define $f \circ h$ to be consistent with the usual definition for functions?

8. Let

$$R_\theta = \begin{pmatrix} \cos\theta & -\sin\theta \\ \sin\theta & \cos\theta \end{pmatrix}$$

denote a rotation in the plane through angle θ. Define $\langle f \circ R_\theta, \varphi \rangle = \langle f, \varphi \circ R_{-\theta} \rangle$. Show that this is consistent with the definition of $f \circ R_\theta$ for functions. If $\langle f, \varphi \rangle = \int_{-\infty}^{\infty} \varphi(x, 0)\, dx$, what is $f \circ R_{\pi/2}$?

9. Say that f is radial if $f \circ R_\theta = f$ for all θ. Is δ radial? Show that $\langle f_R, \varphi \rangle = \frac{1}{2\pi} \int_0^{2\pi} \langle f \circ R_\theta, \varphi \rangle\, d\theta$ defines a radial distribution. Show that f is radial if and only if $f = f_R$.

10. Show that if f is a radial locally integrable function then f is radial as a distribution.

11. Show that if f is a radial distribution then

$$x\frac{\partial f}{\partial y} - y\frac{\partial f}{\partial x} = 0.$$

12. Show that if f is a homogeneous distribution of degree t on $\mathbb{R}^n$ then

$$\sum_{j=1}^{n} x_j \frac{\partial f}{\partial x_j} = tf.$$

13. Show that $\langle \delta^{(k)}, \varphi \rangle = (-1)^k \varphi^{(k)}(0)$ in $\mathbb{R}^1$.

14. In $\mathbb{R}^2$ define $\langle f, \varphi \rangle = \int_{-\infty}^{\infty} \int_0^{\infty} \varphi(x, y)\, dx\, dy$. Compute $\partial f/\partial x$ and $\partial f/\partial y$.

15. In $\mathbb{R}^2$ define $\langle f, \varphi \rangle = \int_0^{\infty} \int_0^{\infty} \varphi(x, y)\, dx\, dy$. Compute $\partial^2 f/\partial x\partial y$.

16. In $\mathbb{R}^1$ compute $m\delta''$ explicitly for m a C^∞ function.

17. Show that every test function $\varphi \in \mathcal{D}(\mathbb{R}^1)$ can be written $\varphi = \psi' + c\varphi_0$ where φ_0 is a fixed test function (with $\int_{-\infty}^{\infty} \varphi_0(x)\, dx \neq 0$), $\psi \in \mathcal{D}$ and c a constant. (*Hint*: Choose $\psi(x) = \int_{-\infty}^{x} (\varphi(t) - c\varphi_0(t))\, dt$ for the appropriate choice of c.) Use this to prove: If f is a distribution on $\mathbb{R}^1$ satisfying $f' = 0$ then f is a constant.

18. Generalize problem *17* to show that if f is a distribution on $\mathbb{R}^1$ satisfying $f^{(k)} = 0$ then f is a polynomial of degree $< k$.

19. Generalize problem *17* to show that if f is a distribution on $\mathbb{R}^n$ satisfying $(\partial/\partial x_k)f = 0$, $k = 1, \ldots, n$ then f is constant.

20. Show that every test function $\varphi \in \mathcal{D}(\mathbb{R}^1)$ can be written $\varphi(x) = x\psi(x) + c\varphi_0(x)$ where φ_0 is a fixed test function (with $\varphi_0(0) \neq 0$), $\psi \in \mathcal{D}$, and c is a constant. Use this to show that if f is a distribution satisfying $xf = 0$, then $f = c\delta$.

21. Generalize problem *20* to show that if f is a distribution on $\mathbb{R}^n$ satisfying $x_j f = 0$, $j = 1, \ldots, n$, then $f = c\delta$.

Chapter 3

Fourier Transforms

3.1 From Fourier series to Fourier integrals

Let's recall the basic facts about Fourier series. Every continuous function $f(x)$ defined on $\mathbb{R}^1$ and periodic of period 2π ($f(x)$) may take real or complex values) has a Fourier series written

$$f(x) \sim \sum_{k=-\infty}^{\infty} a_k e^{ikx}.$$

(We write $\sim$ instead of $=$ because without additional conditions on f, such as differentiability, the series may diverge for particular values of x.) The coefficients in this expansion are given by

$$a_k = \frac{1}{2\pi} \int_{-\pi}^{\pi} f(x) e^{-ikx} dx$$

(the integral may be taken over any interval of length 2π since both $f(x)$ and e^{-ikx} are periodic) and satisfy Parseval's identity

$$\sum_{-\infty}^{\infty} |a_k|^2 = \frac{1}{2\pi} \int_{-\pi}^{\pi} |f(x)|^2 dx.$$

Perhaps you are more familiar with these formulas in terms of sines and cosines:

$$f \sim \frac{b_0}{2} + \sum_{k=1}^{\infty} b_k \cos kx + c_k \sin kx.$$

It is easy to pass back and forth from one to the other using the relation $e^{ikx} = \cos kx + i \sin kx$. Although the second form is a bit more convenient

28

for dealing with real-valued functions (the coefficients b_k and c_k must be real for $f(x)$ to be real, whereas for the coefficients a_k the condition is $a_{-k} = \bar{a}_k$), the exponential form is a lot more simple when it comes to dealing with derivatives.

Returning to the exponential form, we can obtain a Fourier series expansion for functions periodic of arbitrary period T by changing variable. Indeed, if $f(x + T) = f(x)$ then setting

$$F(x) = f\left(\frac{T}{2\pi}x\right)$$

we have

$$F(x + 2\pi) = f\left(\frac{T}{2\pi}(x + 2\pi)\right) = f\left(\frac{T}{2\pi}x + T\right) = f\left(\frac{T}{2\pi}x\right) = F(x)$$

so F is periodic of period 2π. The Fourier series for F is

$$F(x) \sim \sum_{-\infty}^{\infty} a_k e^{ikx}$$

$$a_k = \frac{1}{2\pi}\int_{-\pi}^{\pi} F(x)e^{-ikx}dx$$

$$\sum_{-\infty}^{\infty}|a_k|^2 = \frac{1}{2\pi}\int_{-\pi}^{\pi}|F(x)|^2\,dx.$$

Substituting

$$F(x) = f\left(\frac{T}{2\pi}x\right)$$

and making the change of variable $x \to (2\pi/T)x$ yields

$$f(x) \sim \sum_{-\infty}^{\infty} a_k e^{\frac{i2\pi k}{T}x}$$

$$a_k = \frac{1}{T}\int_{-T/2}^{T/2} f(x)e^{-\frac{i2\pi k}{T}x}dx$$

$$\sum_{-\infty}^{\infty}|a_k|^2 = \frac{1}{T}\int_{T/2}^{T/2}|f(x)|^2\,dx.$$

Now suppose $f(x)$ is not periodic. We still want an analogous expansion. Let's approximate $f(x)$ by restricting it to the interval $-\frac{T}{2} \le x \le \frac{T}{2}$ and extending it periodically of period T. In the end we will let $T \to \infty$. Before passing to the limit let's change notation. Let $\xi_k = 2\pi k / T$ and let $g(\xi_k) = (T/2\pi) a_k$. Then we may rewrite the equations as

$$g(\xi_k) = \frac{1}{2\pi} \int_{-T/2}^{T/2} f(x) e^{-ix\xi_k} dx,$$

$$f(x) \sim \sum_{-\infty}^{\infty} g(\xi_k) e^{ix\xi_k} \frac{2\pi}{T} = \sum_{-\infty}^{\infty} g(\xi_k) e^{ix\xi_k} (\xi_k - \xi_{k-1}),$$

$$\sum_{-\infty}^{\infty} |g(\xi_k)|^2 (\xi_k - \xi_{k-1}) = \frac{1}{2\pi} \int_{-T/2}^{T/2} |f(x)|^2 dx.$$

Written in this way the sums look like approximations to integrals. Notice that the grid of points $\{\xi_k\}$ gets finer as $T \to \infty$, so formally passing to the limit we obtain

$$g(\xi) \quad \sim \quad \frac{1}{2\pi} \int_{-\infty}^{\infty} f(x) e^{-ix\xi} dx$$

$$f(x) \quad \sim \quad \int_{-\infty}^{\infty} g(\xi) e^{ix\xi} dx$$

$$\int_{-\infty}^{\infty} |g(\xi)|^2 d\xi \quad \sim \quad \frac{1}{2\pi} \int_{-\infty}^{\infty} |f(x)|^2 dx$$

where I have written $\sim$ instead of $=$ since we have not justified any of the above steps. But it turns out that we can write $=$ if we impose some conditions on f. The above formulas are referred to as the Fourier transform, the Fourier inversion formula, and the Plancherel formula, although there is some dispute as to which name goes with which formula.

The analogy with Fourier series is clear. Here we have $f(x)$ represented as an integral of the "simple" functions $e^{ix\xi}, \xi \in \mathbb{R}$, instead of a sum of "simple" functions $e^{ikx}, k \in \mathbb{Z}$, and the weighting factor $g(\xi)$ is given by integrating $f(x)$ against the complex conjugate of the "simple" function $e^{ix\xi}$ just as the coefficient a_k was given, only now the integral extends over the whole line instead of over a period. The third formula expresses the mean-square of $f(x)$ in terms of the mean-square of the weighting factors $g(\xi)$.

Notice the symmetry of the first two formulas (in contrast to the asymmetry of Fourier series). We could also regard $g(\xi)$ as the given function, the second formula defining a weighting factor $f(x)$ and the first giving $g(\xi)$ as an integral of "simple" functions $e^{-ix\xi}$; here ξ is the variable and x is a parameter indexing the "simple" functions.

Let us take this point of view and simultaneously interchange the names of some of the variables.

Definition (formal): Given a complex valued function $f(x)$ of a real variable, define the *Fourier transform* of f, denoted $\mathcal{F}f$ or $\hat{f}$ by $\hat{f}(\xi) = \int_{-\infty}^{\infty} f(x)e^{ix\xi}dx$ and the *inverse Fourier transform*, denoted $\mathcal{F}^{-1}f$ or $\check{f}$, by $\check{f}(\xi) = \frac{1}{2\pi}\int_{-\infty}^{\infty} f(x)e^{-ix\xi}dx$ (note that $\check{f}(\xi) = \frac{1}{2\pi}\hat{f}(-\xi)$).

Then the formal derivation given above indicates that we expect

$$\mathcal{F}^{-1}(\mathcal{F}f) = f, \quad \mathcal{F}(\mathcal{F}^{-1}f) = f \quad \text{(Fourier inversion formula)}$$

and

$$\int_{-\infty}^{\infty} |f(x)|^2 dx = \frac{1}{2\pi}\int_{-\infty}^{\infty}|\hat{f}(\xi)|^2 d\xi \quad \text{(Plancherel formula)}.$$

Warning: There is no general agreement on whether to put the minus sign with $\hat{f}$ or $\check{f}$, or on what to do with the $1/2\pi$. Sometimes the $1/2\pi$ is put with $\mathcal{F}^{-1}$, sometimes it is split up so that a factor of $1/\sqrt{2\pi}$ is put with both $\mathcal{F}$ and $\mathcal{F}^{-1}$, and sometimes the 2π crops up in the exponential $\left(\int e^{2\pi ix\xi}f(x)\,dx\right)$, in which case it disappears altogether as a factor. Therefore, in using Fourier transform formulas from different sources, always check which definition is being used. This is a great nuisance—for me as well as for you.

Returning now to our Fourier inversion formula $\mathcal{F}\mathcal{F}^{-1}f = f$ (or $\mathcal{F}^{-1}\mathcal{F}f = f$; they are equivalent if you make a change of variable), we want to know for which functions f it is valid. In the case of Fourier series there was a smoothness condition (differentiability) on f for $f(x) = \Sigma a_k e^{ikx}$ to hold. In this case we will need more, for in order for the improper integral $\int_{-\infty}^{\infty} f(x)e^{ik\xi}dx$ to exist we must have $f(x)$ tending to zero sufficiently rapidly. Thus we need a combination of smoothness and decay at infinity. We now turn to a discussion of a class of functions that has these properties (actually it has much more than is needed for the Fourier inversion theorem).

3.2   The Schwartz class S

Eventually we will want to deal with functions of several variables $f(x) = f(x_1, \ldots, x_n)$. We say such a function is *rapidly decreasing* if there are constants M_n such that

$$|f(x)| \leq M_N |x|^{-N} \text{ as } x \to \infty,$$

for $N = 1, 2, 3, \ldots$.

Another way of saying this is that after multiplication by any polynomial $p(x)$, $p(x)f(x)$ still goes to zero as $x \to \infty$. A C^∞ function is of class $S(\mathbb{R}^n)$ if f and all its partial derivatives are rapidly decreasing. Any function in $\mathcal{D}(\mathbb{R}^n)$ also belongs to $S(\mathbb{R}^n)$, but $S(\mathbb{R}^n)$ is a larger class of functions (it is sometimes called "the Schwartz class"). A typical function in $S(\mathbb{R}^n)$ is $e^{-|x|^2}$; this does not have bounded support so it is not in $\mathcal{D}(\mathbb{R}^n)$. (To verify that $e^{-|x|^2} \in S(\mathbb{R}^n)$, note that any derivative is a polynomial times $e^{-|x|^2}$, and $e^{-|x|^2}$ decreases fast enough as $x \to \infty$ that it beats out the growth of any polynomial.)

We need the following elementary properties of the class $S(\mathbb{R}^n)$:

1. $S(\mathbb{R}^n)$ is a vector space; it is closed under linear combinations.

2. $S(\mathbb{R}^n)$ is an algebra; the product of functions in $S(\mathbb{R}^n)$ also belongs to $S(\mathbb{R}^n)$ (this follows from Leibniz' formula for derivatives of products).

3. $S(\mathbb{R}^n)$ is closed under multiplication by polynomials (this is not a special case of 2) since polynomials are not in S.

4. $S(\mathbb{R}^n)$ is closed under differentiation.

5. $S(\mathbb{R}^n)$ is closed under translations and multiplication by complex exponentials $e^{ix \cdot \xi}$.

6. $S(\mathbb{R}^n)$ functions are integrable: $\int_{\mathbb{R}^n} |f(x)| dx < \infty$ for $f \in S(\mathbb{R}^n)$. This follows from the fact that $|f(x)| \le M(1 + |x|)^{-(n+1)}$ and

$$\int_{\mathbb{R}^n} (1 + |x|)^{-(n+1)} dx = \text{(polar coordinates)}$$

$$C \int_0^\infty (1 + r)^{-n-1} r^{n-1} dr < \infty$$

(decreases like r^{-2} at infinity).

3.3 Properties of the Fourier transform on S

The reason $S(\mathbb{R}^n)$ is useful in studying Fourier transforms is that $\hat{f} \in S(\mathbb{R}^n)$ whenever $f \in S(\mathbb{R}^n)$.

Actually I have only given you the definition of $\hat{f}$ for $\mathbb{R}^1$, but for $\mathbb{R}^n$ the situation is similar:

$$\hat{f}(\xi) = \int_{\mathbb{R}^n} f(x) e^{ix \cdot \xi} dx$$

where $x, \xi \in \mathbb{R}^n, x \cdot \xi = x_1\xi_1 + \cdots + x_n\xi_n$ and

$$\check{f}(\xi) = \frac{1}{(2\pi)^n} \int_{\mathbb{R}^n} f(x)e^{-ix\cdot\xi}dx.$$

It takes a little bit of work to show that $\mathcal{F}$ preserves the class $\mathcal{S}(\mathbb{R}^n)$. In the process we will derive some very important formulas. First observe that the Fourier transform is at least *defined* for $f \in \mathcal{S}(\mathbb{R}^n)$ since the integrand $f(x)e^{ix\cdot\xi}$ has absolute value $|f(x)|$ and this has finite integral. Next we have to see how $\hat{f}$ changes as we do various things to f. As we do this let's keep in mind the symmetry of the Fourier transform and its inverse. If doing "ping" to f does "pong" to $\hat{f}$, then we should expect doing "pong" to f does "ping" to $\hat{f}$, except for those factors of 2π and that minus sign.

1. *Translation:*

$$\tau_y f(x) = f(x+y)$$
$$\mathcal{F}(\tau_y f)(\xi) = \int_{\mathbb{R}^n} f(x+y)e^{ix\cdot\xi}dx = \int_{\mathbb{R}^n} f(x)e^{i(x-y)\cdot\xi}dx$$

(change of variable $x \to x - y$).

But $e^{i(x-y)\cdot\xi} = e^{(ix\cdot\xi - iy\cdot\xi)} = e^{-iy\cdot\xi}e^{ix\cdot\xi}$ and we can take the factor $e^{-iy\cdot\xi}$ outside the integral since it is independent of x. So

$$\mathcal{F}(\tau_y f)(\xi) = e^{-iy\cdot\xi}\mathcal{F}f(\xi)$$

(translation by the vector y goes over to multiplication by the exponential $e^{-iy\cdot\xi}$ on the Fourier transform side).

2. Dually, we expect multiplication by an exponential to go over to translation on the Fourier transform side. Note the minus sign is gone.

$$\mathcal{F}(e^{ix\cdot y}f(x))(\xi) = \int e^{ix\cdot y}f(x)e^{ix\cdot\xi}dx$$
$$= \int f(x)e^{ix\cdot(\xi+y)}dx$$
$$= \mathcal{F}f(\xi+y) = \tau_y\mathcal{F}f(x).$$

3. *Differentiation:*

$$\mathcal{F}\left(\frac{\partial}{\partial x_k}f\right)(\xi) = \int e^{ix\cdot\xi}\frac{\partial f}{\partial x_k}(x)\,dx$$
$$= -\int f(x)\frac{\partial}{\partial x_k}(e^{ix\cdot\xi})\,dx$$
$$= -i\xi_k\int f(x)e^{ix\cdot\xi}\,dx$$
$$= -i\xi_k\mathcal{F}f(\xi)$$

where we have integrated by parts in the x_k variable (the boundary terms are zero because f is rapidly decreasing).

Exercise: Derive the same formula by using 1 above and

$$\frac{\partial f}{\partial x_k} = \lim_{h \to 0} \frac{T_{he_k} f - f}{h}.$$

Thus differentiation goes over to multiplication by a monomial. More generally, by iterating this formula, partial derivatives of higher order go over to multiplication by polynomials. We can write this concisely as

$$\mathcal{F}\left(p\left(\frac{\partial}{\partial x}\right) f\right) = p(-i\xi)\mathcal{F}f(\xi)$$

for any polynomial p, where $p\left(\partial/\partial x\right)$ means we replace x_k by $\partial/\partial x_k$ in the polynomial.

4. Dually, we expect multiplication by a polynomial to go over to differentiation:

$$
\begin{aligned}
\frac{\partial}{\partial \xi_k}\mathcal{F}f(\xi) &= \frac{\partial}{\partial \xi_k}\int f(x)e^{ix\cdot\xi}\,dx \\
&= \int f(x)\frac{\partial}{\partial \xi_k}e^{ik\cdot\xi}\,dx \\
&= \int ix_k f(x)e^{ix\cdot\xi}\,dx = i\mathcal{F}(x_k f(x))
\end{aligned}
$$

(the interchange of the derivative and integral is valid because $f \in S$). Iterating, we obtain

$$\mathcal{F}(p(x)f(x))(\xi) = p\left(-i\frac{\partial}{\partial \xi}\right)\mathcal{F}f(\xi).$$

5. *Convolution:* If $f, g \in S$, define the convolution $f * g$ by

$$f * g(x) = \int f(x - y)g(y)\,dy.$$

By a change of variable $y \to x - y$ this is equal to $\int f(y)g(x - y)\,dy$ so $f * g = g * f$.

Exercise: Verify the associative law for convolutions $f * (g * h) = (f * g) * h$ by similar manipulations.

Because of the rapid decrease of f and g, the convolution integral is defined. It is possible (although tricky) to show that if $f, g \in S$ then $f * g \in S$. One important property of convolutions is that derivatives may be placed on either factor (but not both):

$$
\begin{aligned}
\frac{\partial}{\partial x_k} f * g(x) &= \frac{\partial}{\partial x_k} \int f(x - y) g(y) \, dy \\
&= \int \frac{\partial f}{\partial x_k} (x - y) g(y) \, dy \\
&= \frac{\partial f}{\partial x_k} * g(x)
\end{aligned}
$$

but also

$$
\begin{aligned}
\frac{\partial}{\partial x_k} f * g(x) &= \frac{\partial}{\partial x_k} \int f(y) g(x - y) \, dy \\
&= \int f(y) \frac{\partial g}{\partial x_k} (x - y) \, dy \\
&= f * \frac{\partial g}{\partial x_k} (x).
\end{aligned}
$$

Convolutions are important because solutions to differential equations are often given by convolutions where one factor is the given function and the other is a special "kernel."

Under Fourier transforms the convolution goes over to multiplication. To see this, start with the definition

$$
\mathcal{F}(f * g)(\xi) = \iint f(x - y) g(y) dy \, e^{ix \cdot \xi} \, dx.
$$

Write $e^{ix \cdot \xi} = e^{i(x-y) \cdot \xi} e^{iy \cdot \xi}$ and interchange the order of integration:

$$
\mathcal{F}(f * g)(\xi) = \int \left(\int f(x - y) e^{i(x-y) \cdot \xi} dx \right) \cdot g(y) e^{iy \cdot \xi} \, dy.
$$

For each fixed y make the change of variable $x \to x + y$ in the inner integral and you have just the Fourier transform of f. Thus

$$
\begin{aligned}
\mathcal{F}(f * g)(\xi) &= \int \mathcal{F} f(\xi) g(y) e^{iy \cdot \xi} dy \\
&= \mathcal{F} f(\xi) \mathcal{F} g(\xi).
\end{aligned}
$$

6. Dually, we expect multiplication to go over to convolution. *This cannot be verifed directly* but must use the Fourier inversion formula! We can rewrite $\mathcal{F} f \cdot \mathcal{F} g = \mathcal{F}(f * g)$ as

$$
\mathcal{F}^{-1} f \cdot \mathcal{F}^{-1} g = \frac{1}{(2\pi)^n} \mathcal{F}^{-1}(f * g),
$$

since

$$\mathcal{F}^{-1}F(x) = \frac{1}{(2\pi)^n}\mathcal{F}F(-x).$$

Substituting $\hat{f}$ for f and $\hat{g}$ for g yields

$$f \cdot g = \mathcal{F}^{-1}\hat{f} \cdot \mathcal{F}^{-1}\hat{g} = \frac{1}{(2\pi)^n}\mathcal{F}^{-1}(\hat{f} * \hat{g}).$$

Thus

$$\mathcal{F}(f \cdot g) = \mathcal{F}\left(\frac{1}{(2\pi)^n}\mathcal{F}^{-1}(\hat{f} * \hat{g})\right)$$

$$= \frac{1}{(2\pi)^n}\hat{f} * \hat{g}.$$

Exercise: Use this kind of argument to pass from 1 to 2 above.
We now collect together all these identities in a "ping-pong" table:

Function (x)	Fourier transform (ξ)
$f(x)$	$\hat{f}(\xi)$
$f(x+y)$	$e^{-iy\cdot\xi}\hat{f}(\xi)$
$e^{ix\cdot y}f(x)$	$\hat{f}(\xi+y)$
$\dfrac{\partial f}{\partial x_k}(x)$	$-i\xi_k\hat{f}(\xi)$
$p\left(\dfrac{\partial}{\partial x}\right)f(x)$	$p(-i\xi_k)\hat{f}(\xi)$
$x_k f(x)$	$-i\dfrac{\partial\hat{f}}{\partial\xi_k}(\xi)$
$p(x)f(x)$	$p\left(-i\dfrac{\partial}{\partial\xi}\right)\hat{f}(\xi)$
$f * g(x)$	$\hat{f}(\xi)\hat{g}(\xi)$
$f(x)g(x)$	$\dfrac{1}{(2\pi)^n}\hat{f} * \hat{g}(\xi)$

Many of the above identities are valid for more general classes of functions than $\mathcal{S}$. We will return to this question later.

We should pause now to consider the significance of this table. Some of the operations are simple, and some are complicated. For instance, differentiation is a complicated operation, but the corresponding entry in the table is multiplication by a polynomial, a simple operation. Thus the Fourier transform simplifies operations we want to study. Say we want to solve a partial differential equation $p\left(\partial/\partial x\right)u = f, f$ given. Taking Fourier transform this equation becomes

$$p(-i\xi)\hat{u}(\xi) = \hat{f}(\xi).$$

This is a simple equation to solve for $\hat{u}$; we divide

$$\hat{u}(\xi) = \frac{1}{p(-i\xi)}\hat{f}(\xi).$$

If we want u we invert the Fourier transform

$$u = \mathcal{F}^{-1}(\hat{u}) = \mathcal{F}^{-1}\left(\frac{1}{p(-i\xi)}\hat{f}(\xi)\right).$$

From the table we know

$$\frac{1}{p(-i\xi)}\hat{f}(\xi) = \mathcal{F}(g * f)$$

if

$$\hat{g} = \frac{1}{p(-i\xi)},$$

i.e., if

$$g = \mathcal{F}^{-1}\left(\frac{1}{p(-i\xi)}\right).$$

Thus $u = g * f$. Thus the problem is to compute

$$\mathcal{F}^{-1}\left(\frac{1}{p(-i\xi)}\right).$$

In many cases this can be done explicitly. Of course this is an oversimplification because in general it is difficult to make sense of $\mathcal{F}^{-1}(1/p(-i\xi))$, especially if $p(-i\xi)$ has zeroes for $\xi \in \mathbb{R}^n$. At any rate, this is the basic idea of Fourier analysis—by using the above table many diverse sorts of problems may be reduced to the problem of computing Fourier transforms.

3.4 The Fourier inversion formula on S

Now we return to our goal of showing that if $f \in S$ then $\hat{f} \in S$. We have all the ingredients ready except for the following simple estimate:

$$|\hat{f}(\xi)| \leq \int |f(x)| \, dx$$

whenever f integrable. This is because

$$
\begin{aligned}
|\hat{f}(\xi)| &= \left| \int f(x) e^{ix\cdot\xi} \, dx \right| \\
&\leq \int \left| f(x) e^{ix\cdot\xi} \right| \, dx
\end{aligned}
$$

and $|e^{ix\cdot\xi}| = 1$.

Now let $f \in S$. We want to show that $\hat{f}$ is rapidly decreasing. That means that $p(\xi)\hat{f}(\xi)$ remains bounded for any polynomial p. But by the table $p(\xi)\hat{f}(\xi) = \mathcal{F}(p(i\frac{\partial}{\partial x})f)$ and so

$$|p(\xi)\hat{f}(\xi)| \leq \int \left| p\left(i\frac{\partial}{\partial x}\right) f(x) \right| dx.$$

But

$$p\left(i\frac{\partial}{\partial x}\right) f$$

is rapidly decreasing so the integral is finite. To summarize : if $f \in S$ then $\hat{f}$ is rapidly decreasing. To show $\hat{f} \in S$ we must also show that all derivatives of $\hat{f}$ are also rapidly decreasing. But by the table any derivative of $\hat{f}$ is the Fourier transform of a polynomial times f :

$$p\left(\frac{\partial}{\partial \xi}\right) \hat{f}(\xi) = \mathcal{F}(p(ix)f(x)).$$

But multiplying f by a polynomial produces another function in S: $p(ix)f(x)$ $= g(x) \in S$. By what we have just done $\hat{g}(\xi)$ is rapidly decreasing so $p(\frac{\partial}{\partial \xi})\hat{f}(\xi) = \hat{g}(\xi)$ is rapidly decreasing. Thus $\hat{f} \in S$.

The essence of the above argument is the duality under the Fourier transform of differentiability and decrease at infinity. Roughly speaking, the more derivatives f has, the faster $\hat{f}$ decreases at infinity, while the faster f decreases at infinity the more derivatives $\hat{f}$ has.

More precisely the estimates look like this:

(a) If f has derivatives of orders up to N that are all integrable, then $\hat{f}$ decreases at infinity like $|\xi|^{-N}$.

(b) If f decreases at infinity like $|\xi|^{-N}$ then $\hat{f}$ has continuous and bounded derivatives of all orders up to $N - n - 1$.

The trouble with estimates (a) and (b) is that they are not comparable—there is a loss of $n + 1$ derivatives in (b), and while in (a) the derivatives must be integrable, in (b) they turn out to be just bounded. These defects disappear when we consider the class S because there are an infinite number of derivatives and a decrease at infinity faster than $|\xi|^{-N}$ for any N.

Now that we know $\hat{f} \in S$ whenever $f \in S$ we can be sure that $\mathcal{F}^{-1}\hat{f}$ is defined, and so the Fourier inversion formula $\mathcal{F}^{-1}\hat{f} = f$ (and also $\mathcal{F}\check{f} = f$) is at least plausible. We have already seen a derivation of this formula from Fourier series. At this point I want to give one more explanation of why it is true, based on the idea of "summability," which is important in its own right. For simplicity I will restrict the discussion to the case $n = 1$.

We could attempt to write the Fourier inversion formula as a double integral, namely

$$f(x) = \frac{1}{2\pi} \int_{-\infty}^{\infty} \int_{-\infty}^{\infty} e^{-i(x-y)\cdot\xi} f(y) \, dy \, d\xi.$$

The trouble with this double integral is that it is not absolutely convergent; if you take the absolute value of the integrand you eliminate the oscillating factor $e^{-i(x-y)\cdot\xi}$ and the resultant integral

$$\int_{-\infty}^{\infty} \int_{-\infty}^{\infty} |f(y)| \, dy \, d\xi$$

clearly diverges. This means we cannot interchange the order of integration in the usual sense, although later we will do exactly this in the distribution sense.

Proceeding more cautiously, we throw a Gaussian factor $e^{-t|\xi|^2}$ into the integral:

$$\frac{1}{2\pi} \int_{-\infty}^{\infty} \int_{-\infty}^{\infty} e^{-i(x-y)\cdot\xi} f(y) e^{-t|\xi|^2} \, dy \, d\xi.$$

The result is no longer equal to $f(x)$, of course, but if we let $t \to 0$ the factor $e^{-t|\xi|^2}$ tends to one, so we should obtain $f(x)$ in the limit. Of course nothing has been proved yet. But the double integral is now absolutely convergent and so we can evaluate it by taking the iterated integral in either order. If we take the indicated order we obtain

$$\frac{1}{2\pi} \int_{-\infty}^{\infty} e^{-ix\cdot\xi} \hat{f}(\xi) e^{-t|\xi|^2} \, d\xi,$$

while if we reverse the order of integration we obtain

$$\int_{-\infty}^{\infty} f(y) \left(\frac{1}{2\pi} \int_{-\infty}^{\infty} e^{-i(x-y)\cdot\xi} e^{-t|\xi|^2} d\xi \right) dy$$

$$= \int_{-\infty}^{\infty} f(y) G_t(x-y) \, dy$$

where

$$G_t(x-y) = \frac{1}{2\pi} \int_{-\infty}^{\infty} e^{-i(x-y)\cdot\xi} e^{-t|\xi|^2} d\xi,$$

which just means that G_t is the inverse Fourier transform of the Gaussian function $e^{-t|\xi|^2}$.

In a moment we will compute $G_t(x) = \frac{1}{\sqrt{4\pi t}} e^{-|x|^2/4t}$. Granted this, we have

$$\frac{1}{2\pi} \int_{-\infty}^{\infty} e^{-ix\cdot\xi} \hat{f}(\xi) e^{-t|\xi|^2} d\xi = \int_{-\infty}^{\infty} f(y) G_t(x-y) \, dy$$

for all $t > 0$, and we take the limit as $t \to 0$. The limit of the left side is $\frac{1}{2\pi} \int e^{-ix\cdot\xi} \hat{f}(\xi) \, d\xi$, which is just $\mathcal{F}^{-1}\mathcal{F}f(x)$. The limit on the right side is $f(x)$ because it is what is called a convolution with an approximate identity, $G_t * f(x)$. The properties of G_t that make it on approximate identity are the following:

1. $\int_{-\infty}^{\infty} G_t(x)dx = 1$ for all t

2. $G_t(x) \geq 0$

3. $\lim_{t\to 0} G_t(x) = 0$ if $x \neq 0$, uniformly in $|x| \geq \epsilon$ for any $\epsilon > 0$.

Indeed *2* and *3* are obvious, and *1* follows by a change of variable from the formula

$$\int_{-\infty}^{\infty} e^{-x^2} dx = \sqrt{\pi}$$

which we will verify in the process of computing the Fourier transform of the Gaussian.

Thus the Fourier inversion formula is a direct consequence of the approximate identity theorem: if G_t is an approximate identity then $G_t * f \to f$ as $t \to 0$ for any $f \in S$. To see why this theorem is plausible we have to observe that the convolution $G_t * f$ can be interpreted as a weighted average of f (by properties *1* and *2*, with $G_t(x-y)$ as the weighting factor (here x is fixed and y is the variable of integration). But by property *3* this weighting factor is concentrated around $y = x$, and this becomes more pronounced as $t \to 0$, so that in the limit we get $f(x)$.

3.5   The Fourier transform of a Gaussian

Now we compute the Fourier transform and inverse Fourier transform of a Gaussian. We can do the computation in n dimensions, and in fact there is an advantage to allowing general n. You see, the Gaussian function $e^{-t|x|^2}$ (with $t > 0$—we may even allow it to be complex provided $\operatorname{Re} t > 0$) is a function in $S(\mathbb{R}^n)$ that is radial (it depends only on $|x|$) and is a product of one-dimensional functions

$$e^{-t|x|^2} = e^{-tx_1^2} e^{-tx_2^2} \ldots e^{-tx_n^2}.$$

It is possible to show that these two properties essentially characterize this function (every radial function in $S(\mathbb{R}^n)$ that is a product of one-dimensional functions is of the form $ce^{-t|x|^2}$ for some constant c and complex t with $\operatorname{Re} t > 0$). Since the Fourier transform preserves these properties we expect the Fourier transform to have the same form. But we give a more direct computation.

The method used is the calculus of residues. We have

$$
\begin{aligned}
f(x) &= e^{-t|x|^2} \\
\hat{f}(\xi) &= \int_{\mathbb{R}^n} e^{-t|x|^2} e^{ix\cdot\xi}\, dx \\
&= \int_{-\infty}^{\infty} \ldots \int_{-\infty}^{\infty} e^{-tx_1^2 + ix_1\xi_1} e^{-tx_2^2 + ix_2\xi_2} \ldots e^{-tx_n^2 + ix_n\xi_n}\, dx_1 \ldots dx_n \\
&= \left(\int_{-\infty}^{\infty} e^{-tx_1^2 + ix_1\xi_1}\, dx_1 \right) \left(\int_{-\infty}^{\infty} e^{-tx_2^2 + ix_2\xi_2}\, dx_2 \right) \\
&\quad \ldots \left(\int_{-\infty}^{\infty} e^{-tx_n^2 + ix_n\xi_n}\, dx_n \right)
\end{aligned}
$$

so the problem is reduced to a one-dimensional Fourier transform

$$\int_{-\infty}^{\infty} e^{-tx^2 + ix\xi}\, dx.$$

Now the exponent $-tx^2 + ix\xi$ is a quadratic polynomial in x, so it is natural to "complete the square":

$$-tx^2 + ix\xi = -t\left(x - \frac{i\xi}{2t} \right)^2 - \frac{\xi^2}{4t}$$

so

$$\int_{-\infty}^{\infty} e^{-tx^2 + ix\xi}\, dx = e^{-\xi^2/4t} \int_{-\infty}^{\infty} e^{-t(x - i\xi/2t)^2}\, dx.$$

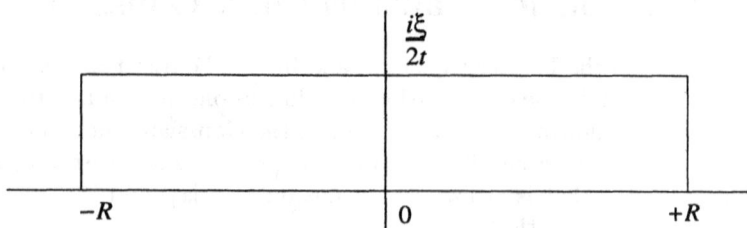

Figure 3.1

Now it is tempting to make the change of variable $x \to x + i\xi/2t$ in this last integral to eliminate the dependence on ξ. But this is a complex change of variable and must be justified by a contour integral. Consider the contour C_R in the following figure for $\xi < 0$ (if $\xi > 0$ the rectangle will lie below the x-axis). Since e^{-tz^2} is analytic,

$$\int_{C_R} e^{-tz^2} dz = 0$$

by the Cauchy integral theorem. The integral along the x-axis side is

$$\int_{-R}^{R} e^{-tx^2} dx.$$

The integral along the parallel side is

$$-\int_{-R}^{R} e^{-t(x - i\xi/2t)^2} dx,$$

the minus sign coming from the fact that along C_R this side is traced in the negative direction. The integrals along the vertical sides tend to zero as $R \to \infty$: $e^{-tz^2} = e^{-t(x^2 - y^2) - 2itxy}$ so $|e^{-tz^2}| \le e^{-t(x^2 - y^2)}$; on the vertical sides $e^{-tx^2} = e^{-tR^2}$, which tends to zero while e^{ty^2} is at most $e^{t(\xi/2t)^2}$ which is bounded independent of R, so the absolute value of the integrand tends to zero and the length of the sides is fixed.

Thus as $R \to \infty$ we get

$$\int_{-\infty}^{\infty} e^{-tx^2} dx - \int_{-\infty}^{\infty} e^{-t(x - i\xi/2t)^2} dx = 0.$$

Our computation will be complete once we evaluate $g(t) = \int_{-\infty}^{\infty} e^{-tx^2} dx$.

There is a famous trick for doing this:

$$
\begin{aligned}
g(t)^2 &= \int_{-\infty}^{\infty} e^{-tx^2}\,dx \cdot \int_{-\infty}^{\infty} e^{-ty^2}\,dy \\
&= \int_{-\infty}^{\infty} \int_{-\infty}^{\infty} e^{-t(x^2+y^2)}\,dx\,dy.
\end{aligned}
$$

Switching to polar coordinates,

$$
\begin{aligned}
g(t)^2 &= \int_0^{2\pi} \int_0^{\infty} e^{-tr^2} r\,dr\,d\theta \\
&= 2\pi \int_0^{\infty} e^{-tr^2} r\,dr \\
&= -2\pi \left. \frac{e^{-tr^2}}{2t} \right]_0^{\infty} = \frac{\pi}{t}.
\end{aligned}
$$

so $g(t) = \sqrt{\pi/t}$. Thus we have

$$
\int_{-\infty}^{\infty} e^{-tx^2} e^{ix\xi}\,dx = \sqrt{\pi/t}\, e^{-\xi^2/4t}
$$

and so

$$
\int_{\mathbb{R}^n} e^{-t|x|^2} e^{ix\cdot\xi}\,dx = \left(\frac{\pi}{t}\right)^{n/2} e^{-|\xi|^2/4t},
$$

and

$$
\frac{1}{(2\pi)^n} \int_{\mathbb{R}^n} e^{-t|\xi|^2} e^{-ix\cdot\xi}\,d\xi = \frac{1}{(4\pi t)^{n/2}} e^{-|x|^2/4t}.
$$

An interesting special case is when $t = 1/2$; then if $f(x) = e^{-|x|^2/2}$ we have $\mathcal{F}f = (2\pi)^{n/2} f$.

This formula is the starting point for many Fourier transform computations, and it is useful for solving the heat equation and Schrödinger's equation.

3.6 Problems

1. Compute $e^{-s|x|^2} * e^{-t|x|^2}$ using the Fourier inversion formula.

2. Show that $\mathcal{F}(d_r f) = r^{-n} d_{1/r} \mathcal{F} f$ for $r > 0$ and any $f \in \mathcal{S}$ where $d_r f(x) = f(rx_1, \ldots, rx_n)$.

3. Show that $f(x)$ is real valued if and only if $\hat{f}(-\xi) = \overline{\hat{f}(\xi)}$.

4. Compute $\mathcal{F}(xe^{-tx^2})$ in $\mathbb{R}^1$.

5. Compute $\mathcal{F}(e^{-ax^2+bx+c})$ in $\mathbb{R}^1$ for $a > 0$. (*Hint*: Complete the square and use the ping-pong table.)

6. Let
$$f(x) = \begin{cases} 1 & \text{if } -a < x < a \\ 0 & \text{otherwise} \end{cases}$$
in $\mathbb{R}^1$. Compute $\mathcal{F}f$ (note f is not in $\mathcal{S}$ but it is integrable).

7. Compute
$$\mathcal{F}\left(\left(\frac{d^2}{dx^2} - x^2\right)f\right)$$
in $\mathbb{R}^1$ in terms of $\mathcal{F}f$.

8. Let
$$R_\theta = \begin{pmatrix} \cos\theta & -\sin\theta \\ \sin\theta & \cos\theta \end{pmatrix}$$
be the rotation in $\mathbb{R}^2$ through angle θ. Show $\mathcal{F}(f \circ R_\theta) = (\mathcal{F}f) \circ R_\theta$. Use this to show that the Fourier transform of a radial function is radial.

9. If $f \in S(\mathbb{R}^1)$ is even, show that $f(|x|) \in S(\mathbb{R}^n)$.

10. Show that $e^{-|x|^a}$ is in $S(\mathbb{R}^1)$ if and only if a is an even positive integer.

11. Show that $\int_{\mathbb{R}^n} f * g(x)\,dx = \int_{\mathbb{R}^n} f(x)\,dx \cdot \int_{\mathbb{R}^n} g(x)\,dx$.

12. Compute the moments $\int_{-\infty}^{\infty} x^k f(x)\,dx$ of a function $f \in S(\mathbb{R}^1)$ in terms of the Fourier transform $\hat{f}$.

13. Show that $\mathcal{F}\tilde{f} = \widetilde{\mathcal{F}f}$ for $\tilde{f}(x) = f(-x)$.

14. Show that $\mathcal{F}^4 f = (2\pi)^{2n} f$. What does this say about possible eigenvalues λ for $\mathcal{F}$ (i.e., $\mathcal{F}f = \lambda f$ for nonzero $f \in S(\mathbb{R}^n)$)?

15. Show that $\left(\frac{d}{dt} + x\right)f = 0$ has $ce^{-x^2/2}$ as its solution. Using the uniqueness theorem for ordinary differential equations, we can conclude that there are no other solutions. What does this tell us about $\mathcal{F}(e^{-x^2/2})$?

16. What is the Fourier transform of $\left(\frac{d}{dx} - x\right)^k e^{-x^2/2}$?

17. Let $f \in S(\mathbb{R}^1)$ and let

$$g(x) = \sum_{k=-\infty}^{\infty} f(x + 2\pi k).$$

Then g is periodic of period 2π. What is the relationship between the Fourier series of g and $\hat{f}$?

18. Show that $\int_{-\infty}^{\infty} x^2 |f(x)|^2 dx = \frac{1}{2\pi} \int_{-\infty}^{\infty} |\hat{f}'(\xi)|^2 d\xi$ and $\int_{-\infty}^{\infty} |f'(x)|^2 dx = \frac{1}{2\pi} \int_{-\infty}^{\infty} \xi^2 |\hat{f}(\xi)|^2 d\xi$.

19. Show that $\int_{\mathbb{R}^n} f(x)\overline{g(x)}\, dx = \frac{1}{(2\pi)^n} \int_{\mathbb{R}^n} \hat{f}(\xi)\overline{\hat{g}(\xi)}\, d\xi$.

20. If $f \in S$ is positive, show that $|\hat{f}(\xi)|^2$ achieves its maximum value at $\xi = 0$.

21. Let $\varphi(x) = f(x, 0)$ for $f \in S(\mathbb{R}^2)$. What is the relationship between $\hat{\varphi}$ and $\hat{f}$?

Chapter 4

Fourier Transforms of Tempered Distributions

4.1 The definitions

You may already have noticed a similarity between the spaces S and $\mathcal{D}$. Since distributions were defined to be linear functionals on $\mathcal{D}$, it seems plausible that linear functionals on S should be of interest. They are, and they are called *tempered distributions*. As the nomenclature suggests, the class of tempered distributions (denoted $S'(\mathbb{R}^n)$) should be a subclass of the distributions $\mathcal{D}'(\mathbb{R}^n)$. This is in fact the case: any f in $S'(\mathbb{R}^n)$ is a functional $\langle f, \varphi \rangle$ defined for all $\varphi \in S(\mathbb{R}^n)$. But since $\mathcal{D}(\mathbb{R}^n) \subseteq S(\mathbb{R}^n)$ this defines by restriction a functional on $\mathcal{D}(\mathbb{R}^n)$, hence a distribution in $\mathcal{D}'(\mathbb{R}^n)$ (it is also true that the continuity of f on S implies the continuity of f on $\mathcal{D}$, but I have not defined these concepts yet). Different functionals on S define different functionals on $\mathcal{D}$ (in other words $\langle f, \varphi \rangle$ is completely determined if you know it for $\varphi \in \mathcal{D}$) so we will be sloppy and not make any distinction between a tempered distribution and the associated distribution in $\mathcal{D}'(\mathbb{R}^n)$. We are thus thinking of $S'(\mathbb{R}^n) \subseteq \mathcal{D}'(\mathbb{R}^n)$. What is *not* true is that every distribution in $\mathcal{D}'(\mathbb{R}^n)$ corresponds to a tempered distribution. For example, the function e^{x^2} on $\mathbb{R}^1$ defines a distribution

$$\langle f, \varphi \rangle = \int_{-\infty}^{\infty} e^{x^2} \varphi(x) \, dx$$

46

(this is finite because the support of φ is bounded). But $e^{-x^2/2} \in \mathcal{S}(\mathbb{R}^1)$ and we would have

$$
\begin{aligned}
\langle f, \varphi \rangle &= \int_{-\infty}^{\infty} e^{x^2} e^{-x^2/2} \, dx \\
&= \int_{-\infty}^{\infty} e^{x^2/2} \, dx = +\infty
\end{aligned}
$$

so there is no way to define f as a tempered distribution. (In fact, it can be shown that a locally integrable function defines a tempered distribution if

$$
\int_{|x| \leq A} |f(x)| \, dx \leq cA^N \text{ as } A \to \infty
$$

for some constants c and N, and this condition is necessary if f is positive.)

Exercise: Verify that if f satisfies this estimate then $\int_{\mathbb{R}^n} |f(x)\varphi(x)| \, dx < \infty$ for all $\varphi \in \mathcal{S}(\mathbb{R}^n)$ so $\int_{\mathbb{R}^n} f(x)\varphi(x) \, dx$ is a tempered distribution.

Now why complicate things by introducing tempered distributions? The answer is that it is possible to define the Fourier transform of a tempered distribution as a tempered distribution, but it is impossible to define the Fourier transform of all distributions in $\mathcal{D}'(\mathbb{R}^n)$ as distributions.

Recall that we were able to define operations on distributions via adjoint identities. If T and S were linear operations that took functions in $\mathcal{D}$ to functions in $\mathcal{D}$ such that

$$
\int (T\psi(x))\varphi(x) \, dx = \int \psi(x) S\varphi(x) \, dx
$$

for $\psi, \varphi \in \mathcal{D}$ we defined

$$
\langle Tf, \varphi \rangle = \langle f, S\varphi \rangle
$$

for any distribution $f \in \mathcal{D}'$. The same idea works for tempered distributions. The adjoint identity for $\psi, \varphi \in \mathcal{S}$ is usually no more difficult than for $\psi, \varphi \in \mathcal{D}$. The only new twist is that the operations T and S must preserve the class $\mathcal{S}$ instead of $\mathcal{D}$. This is true for the operations we discussed previously with one exception: Multiplication by a C^∞ function $m(x)$ is allowed only if $m(x)$ does not grow too fast at infinity; specifically, we require $m(x) \leq c|x|^N$ as $x \to \infty$ for some c and N. This includes polynomials but excludes $e^{|x|^2}$, for $e^{|x|^2} e^{-|x|^2/2}$ is not in $\mathcal{S}$ while $e^{-|x|^2/2} \in \mathcal{S}$.

But in dealing with the Fourier transform it is a real boon to have the class $\mathcal{S}$: If $\varphi \in \mathcal{S}$ then $\mathcal{F}\varphi \in \mathcal{S}$, while if $\varphi \in \mathcal{D}$ it may not be true that $\mathcal{F}\varphi \in \mathcal{D}$ (surprisingly it turns out that if both φ and $\mathcal{F}\varphi$ are in $\mathcal{D}$ then

$\varphi = 0$!!) So all that remains is to discover an adjoint identity involving $\mathcal{F}$. Such an identity should look like

$$\int \hat{\psi}(x)\varphi(x)\,dx = \int \psi(x)S\varphi(x)\,dx$$

where S is an as-yet-to-be-discovered operation.

To get such an identity we substitute the definition $\hat{\psi}(x) = \int \psi(y)e^{ix\cdot y}dy$ and interchange the order of integration

$$\begin{aligned} \int \hat{\psi}(x)\varphi(x)\,dx &= \int\int \psi(y)e^{ix\cdot y}dy\,\varphi(x)\,dx \\ &= \int \psi(y)\left(\int \varphi(x)e^{ix\cdot y}dx\right)dy \\ &= \int \psi(y)\hat{\varphi}(y)\,dy. \end{aligned}$$

We may rename the variable y to obtain

$$\int \hat{\psi}(x)\varphi(x)\,dx = \int \psi(x)\hat{\varphi}(x)\,dx.$$

This is our adjoint identity.

In passing let us note that the Plancherel formula is a simple consequence of this identity. Just take $\psi(x) = \overline{\hat{\varphi}(x)}$. We have $\psi(x) = \int \overline{\varphi(y)}e^{-ix\cdot y}dy = (2\pi)^n \mathcal{F}^{-1}(\overline{\varphi})(x)$. Then $\hat{\psi}(x) = (2\pi)^n \mathcal{F}\mathcal{F}^{-1}\overline{\varphi} = (2\pi)^n \overline{\varphi(x)}$, so the adjoint identity reads

$$(2\pi)^n \int |\varphi(x)|^2 dx = \int |\hat{\varphi}(x)|^2 dx.$$

Now the adjoint identity allows us to define the Fourier transform of a tempered distribution $f \in \mathcal{S}'(\mathbb{R}^n)$: $\langle \hat{f}, \varphi \rangle = \langle f, \hat{\varphi} \rangle$. In other words, $\hat{f}$ is that functional on $\mathcal{S}$ that assigns to φ the value $\langle f, \hat{\varphi} \rangle$. If f is actually a function in $\mathcal{S}$ then $\hat{f}$ is the tempered distribution identified with the function $\hat{f}(x)$. In other words, this definition is consistent with the previous definition, since we are identifying functions $f(x)$ with the distribution

$$\langle f, \varphi \rangle = \int f(x)\varphi(x)\,dx.$$

In fact, more is true. If f is any integrable function we could define the Fourier transform of f directly:

$$\hat{f}(\xi) = \int f(x)e^{ix\cdot\xi}\,dx.$$

Now $\hat{f}$ is a bounded continuous function, so both f and $\hat{f}$ define tempered distributions. The adjoint identity continues to hold:

$$\int \hat{f}(x)\varphi(x)\,dx = \int f(x)\hat{\varphi}(x)\,dx$$

for $\varphi \in \mathcal{S}$ so that $\hat{f}$ is the distribution Fourier transform of f.

The Fourier inversion formula for tempered distributions takes the same form as for functions in $\mathcal{S}$: $\mathcal{F}^{-1}\mathcal{F}f = f$ and $\mathcal{F}\mathcal{F}^{-1}f = f$ with $\mathcal{F}^{-1}f = (2\pi)^{-n}(\mathcal{F}f)\tilde{}$ where the operation $\tilde{f}$ for distributions corresponds to $\tilde{f}(x) = f(-x)$ for functions and is defined by $\langle \tilde{f}, \varphi \rangle = \langle f, \tilde{\varphi} \rangle$. To establish the Fourier inversion formula we just do some definition chasing: since $\varphi = \mathcal{F}\mathcal{F}^{-1}\varphi$ for $\varphi \in \mathcal{S}$ we have

$$
\begin{aligned}
\langle f, \varphi \rangle &= \langle f, \mathcal{F}\mathcal{F}^{-1}\varphi \rangle = \langle \mathcal{F}f, \mathcal{F}^{-1}\varphi \rangle \\
&= (2\pi)^{-n}\langle \mathcal{F}f, (\mathcal{F}\varphi)\tilde{} \rangle = (2\pi)^{-n}\langle (\mathcal{F}f)\tilde{}, \mathcal{F}\varphi \rangle \\
&= \langle \mathcal{F}^{-1}f, \mathcal{F}\varphi \rangle = \langle \mathcal{F}\mathcal{F}^{-1}f, \varphi \rangle.
\end{aligned}
$$

So $\mathcal{F}\mathcal{F}^{-1}f = f$, and similarly for the inversion in the reverse order.

4.2   Examples

1. Let $f = \delta$. What is $\hat{f}$? We must have $\hat{\varphi}(0) = \langle \delta, \hat{\varphi} \rangle = \langle \hat{f}, \varphi \rangle$. But by definition $\hat{\varphi}(0) = \int \varphi(x)\,dx$ so $\hat{f} \equiv 1$. In this example f is not at all smooth, so $\hat{f}$ has no decay at infinity. But f has rapid decay at infinity so $\hat{f}$ is smooth.

2. Let $f = \delta'$ ($n = 1$). Since $f = (d/dx)\delta$ and $\hat{\delta} = 1$ we would like to use our "ping-pong" table to say $\hat{f}(\xi) = -i\xi \cdot \hat{\delta}(\xi) = -i\xi$. This is possible— in fact, the entire table is essentially valid for tempered distributions (for convolutions and products one factor must be in $\mathcal{S}$). Let us verify for instance that

$$\mathcal{F}\left(\frac{\partial}{\partial x_k}f\right) = (-ix_k)\hat{f}$$

for any $f \in \mathcal{S}(\mathbb{R}^n)$. By definition

$$\left\langle \mathcal{F}\left(\frac{\partial}{\partial x_k}f\right), \varphi \right\rangle = \left\langle \frac{\partial}{\partial x_k}f, \hat{\varphi} \right\rangle.$$

By definition

$$\left\langle \frac{\partial}{\partial x_k}f, \hat{\varphi} \right\rangle = -\left\langle f, \frac{\partial}{\partial x_k}\hat{\varphi} \right\rangle.$$

But $\hat{\varphi} \in S$ so by our table $(\partial/\partial x_k)\hat{\varphi} = \mathcal{F}(ix_k\varphi)$. Thus

$$\left\langle \mathcal{F}\left(\frac{\partial}{\partial x_k}f\right), \varphi \right\rangle = -\langle f, \mathcal{F}(ix_k\varphi)\rangle.$$

Now use the definitions of $\mathcal{F}f$:

$$-\langle f, \mathcal{F}(ix_k\varphi)\rangle = -\langle \hat{f}, ix_k\varphi\rangle.$$

Finally, use the definition of multiplication by $-ix_k$:

$$-\langle \hat{f}, ix_k\varphi\rangle = \langle -ix_k\hat{f}, \varphi\rangle.$$

Altogether then,

$$\left\langle \mathcal{F}\left(\frac{\partial}{\partial x_k}f\right), \varphi \right\rangle = \langle -ix_k\hat{f}, \varphi\rangle$$

which is to say $\mathcal{F}(\frac{\partial}{\partial x_k}f) = -ix_k\hat{f}$. The other entries in the table are verified by similar "definition chasing" arguments. Note that δ' is somewhat "rougher" than δ, so its Fourier transform grows at infinity.

3. Let $f(x) = e^{is|x|^2}$, $s \neq 0$ real. Then f is a bounded continuous function, hence it defines a tempered distribution, although f is not integrable so that $\int f(x)e^{ix\cdot\xi}dx$ is not defined.

   In this example the definition of $\hat{f}$ is not very helpful. To be honest, you almost never compute $\hat{f}$ from the definition—instead you use some other method to find out what $\hat{f}$ is and then go back and show it satisfies $\langle \hat{f}, \varphi\rangle = \langle f, \hat{\varphi}\rangle$. That is what we will do in this case.

   Recall that we computed $(e^{-t|\xi|^2})^\wedge(\xi) = (\pi/t)^{n/2} e^{-|\xi|^2/4t}$. We would like to substitute $t = -is$. But note that there is an ambiguity when n is odd, namely which square root to take for $-\pi/is^{n/2}$. This can be clarified by thinking of t as a complex variable z. We must keep Re $z \geq 0$ in order that $e^{-z|x|^2}$ not grow too fast at infinity. But for Re $z \geq 0$ we can determine the square root $z^{1/2}$ uniquely by requiring arg z to satisfy $-\pi/2 \leq$ arg $z \leq \pi/2$. This is consistent with taking the positive square root when z is real and positive. So $-\pi/is = (\pi i/s)$ becomes $e^{\pi i/2}(\pi/|s|)$ when $s > 0$ and $e^{-\pi i/2}(\pi/|s|)$ when $s < 0$ so

$$\left(\frac{\pi}{-is}\right)^{n/2} = \begin{cases} \left(\dfrac{\pi}{|s|}\right)^{n/2} e^{\pi n i/4} & s > 0 \\[2em] \left(\dfrac{\pi}{|s|}\right)^{n/2} e^{-\pi n i/4} & s < 0. \end{cases}$$

With this choice we expect $(e^{is|x|^2})\,\hat{} \ = -\pi/is^{n/2}e^{-i|x|^2/4s}$.

Having first obtained the answer, how do we justify it from the definition? We have to show

$$\langle e^{is|x|^2}, \hat{\varphi}\rangle = (\pi/-is)^{n/2}\langle e^{-i|x|^2/4s}, \varphi\rangle$$

which is to say

$$\int e^{is|x|^2}\hat{\varphi}(x)dx = \left(\frac{\pi}{-is}\right)^{n/2}\int e^{-i|x|^2/4s}\varphi(x)\,dx$$

for all $\varphi \in S$, both integrals being well defined. Now our starting point was
the fact that $\mathcal{F}(e^{-t|x|^2}) = \left(\frac{\pi}{t}\right)^{n/2}e^{-|x|^2/4t}$, which via the adjoint identity
gives

$$\int e^{-t|x|^2}\hat{\varphi}(x)\,dx = \left(\frac{\pi}{t}\right)^{n/2}\int e^{-|x|^2/4t}\varphi(x)\,dx.$$

Now the substitution $t = -is$ may be accomplished by analytic continuation. We consider

$$F(z) = \int e^{-z|x|^2}\hat{\varphi}(x)\,dx$$

and

$$G(z) = \left(\frac{\pi}{z}\right)^{n/2}\int e^{-|x|^2/4z}\varphi(x)\,dx$$

for fixed $\varphi \in S$. For Re $z > 0$ the integrals converge (note that $1/z$ also has
real part > 0) and can be differentiated with respect to z so they define
analytic functions in Re $z > 0$.

We have seen that F and G are equal if z is real (and > 0). But an
analytic function is determined by its values for z real so $F(z) = G(z)$ in
Re $z > 0$. Finally F and G are continuous up to the boundary $z = is$ for
$s \neq 0$,

$$F(is) = \lim_{\epsilon \to 0^+} F(\epsilon + is)$$

and similarly for G (this requires some justification since we are interchanging a limit and an integral), hence $F(is) = G(is)$ which is the result we are
after.

This example illustrates a very powerful method for computing Fourier
transforms via analytic continuation. It has to be used with care, however.
The question of when you can interchange limits and integrals is not just
an academic matter—it can lead to errors if it is misused.

4. Let $f(x) = e^{-t|x|}, t > 0$. This is a rapidly decreasing function but it is
not in S because it fails to be differentiable at $x = 0$. For $n = 1$ it is easy

to compute the Fourier transform directly:

$$
\begin{aligned}
\hat{f}(\xi) &= \int_{-\infty}^{\infty} e^{-t|x|} e^{ix\xi} \, dx \\
&= \int_{-\infty}^{0} e^{tx+ix\xi} \, dx + \int_{0}^{\infty} e^{-tx+ix\xi} \, dx \\
&= \left. \frac{e^{x(t+i\xi)}}{t+i\xi} \right]_{-\infty}^{0} + \left. \frac{e^{x(-t+i\xi)}}{-t+i\xi} \right]_{0}^{\infty} \\
&= \frac{1}{t+i\xi} - \frac{1}{-t+i\xi} = \frac{2t}{t^2+\xi^2}.
\end{aligned}
$$

From the Fourier inversion formula,

$$
e^{-t|x|} = \frac{1}{\pi} \int_{-\infty}^{\infty} \frac{t}{t^2+\xi^2} e^{-ix\xi} \, d\xi.
$$

Exercise: Verify this directly using the calculus of residues.

For $n > 1$ we will use another general method which in outline goes as follows: We try to write $e^{-t|x|}$ as an "average" of Gaussians $e^{-s|x|^2}$. In other words we try to find an identity of the form

$$
e^{-t|x|} = \int_{0}^{\infty} g(s) e^{-s|x|^2} \, ds \quad (g \text{ depends on } t).
$$

If we can do this then (reasoning formally) we should have

$$
\begin{aligned}
\mathcal{F}(e^{-t|x|}) &= \int_{0}^{\infty} g(s) \mathcal{F}(e^{-s|x|^2}) \, ds \\
&= \int_{0}^{\infty} g(s) \left(\frac{\pi}{s}\right)^{n/2} e^{-|x|^2/4s} \, ds.
\end{aligned}
$$

We then try to evaluate this integral (even if we cannot evaluate it explicitly, it may give more information than the original Fourier transform formula).

Now the identity we seek is independent of the dimension because all that appears in it is $|x|$ which is just a positive number (call it λ for emphasis). So we want

$$
e^{-t\lambda} = \int_{0}^{\infty} g(s) e^{-s\lambda^2} \, ds
$$

for all $\lambda > 0$. We will obtain such an identity from the one-dimensional Fourier transform we just computed. We begin by computing

$$
\begin{aligned}
\int_{0}^{\infty} e^{-st^2} e^{-s\xi^2} \, ds &= \left. \frac{e^{-s(t^2+\xi^2)}}{-(t^2+\xi^2)} \right]_{0}^{\infty} \\
&= \frac{1}{t^2+\xi^2}.
\end{aligned}
$$

Since we know $e^{-t|x|} = \frac{1}{\pi} \int_{-\infty}^{\infty} \frac{t}{t^2+\xi^2} e^{-ix\xi} d\xi$ we may substitute in for $1/(t^2 + \xi^2)$ and get

$$e^{-t|x|} = \frac{1}{\pi} \int_{-\infty}^{\infty} t \int_{0}^{\infty} e^{-st^2} e^{-s\xi^2} ds\, e^{-ix\xi} d\xi.$$

Now if we do the ξ-integration first we have

$$\int_{-\infty}^{\infty} e^{-s\xi^2} e^{-ix\xi} d\xi = \left(\frac{\pi}{s}\right)^{1/2} e^{-x^2/4s}$$

so

$$e^{-t|x|} = \int_{0}^{\infty} \frac{t}{(\pi s)^{1/2}} e^{-st^2} e^{-x^2/4s} ds.$$

Putting in λ for $|x|$ we have

$$e^{-t\lambda} = \int_{0}^{\infty} \frac{t}{(\pi s)^{1/2}} e^{-st^2} e^{-\lambda^2/4s} ds$$

which is essentially what we wanted. (We could make the change of variable $s \to 1/4s$ to obtain the exact form discussed above, but this is unnecessary.)
   Now we let x vary in $\mathbb{R}^n$ and substitute $|x|$ for λ to obtain

$$e^{-t|x|} = \int_{0}^{\infty} \frac{t}{(\pi s)^{1/2}} e^{-st^2} e^{-|x|^2/4s} ds$$

so

$$\begin{aligned} \mathcal{F}(e^{-t|x|}) &= \int_{0}^{\infty} \frac{t}{(\pi s)^{1/2}} e^{-st^2} \mathcal{F}(e^{-|x|^2/4s}) ds \\ &= \int_{0}^{\infty} \frac{t}{(\pi s)^{1/2}} e^{-st^2} (4\pi s)^{n/2} e^{-s|\xi|^2} ds. \end{aligned}$$

The last step is to evaluate this integral. We first try to remove the dependence on ξ. Note that $e^{-st^2} e^{-s|\xi|^2} = e^{-s(t^2+|\xi|^2)}$. This suggests the change of variable $s \to s/(t^2 + |\xi|^2)$. Doing this we get

$$\mathcal{F}(e^{-t|x|})(\xi) = \frac{t}{(t^2 + |\xi|^2)^{\frac{n+1}{2}}} \int_{0}^{\infty} \frac{1}{(\pi s)^{1/2}} (4\pi s)^{n/2} e^{-s} ds.$$

This last integral is just a constant depending on n, but not on t or ξ. It can be evaluated in terms of the Γ-function as $2^n \pi^{\frac{n-1}{2}} \Gamma(\frac{n+1}{2})$ (when n is odd $\Gamma(\frac{n+1}{2}) = (\frac{n+1}{2})!$ while if n is even $\Gamma(\frac{n+1}{2}) = \frac{n-1}{2} \cdot \frac{n-3}{2} \dots \frac{1}{2}\sqrt{\pi}$). Thus

$$\mathcal{F}(e^{-t|x|})(\xi) = 2^n \pi^{\frac{n-1}{2}} \Gamma\left(\frac{n+1}{2}\right) \frac{t}{(t^2 + |\xi|^2)^{\frac{n+1}{2}}}.$$

Note this agrees with our previous computation where $n = 1$. Once again we see the decay at infinity of $e^{-t|x|}$ mirrored in the smoothness of its Fourier transform, while the lack of smoothness of $e^{-t|x|}$ at $x = 0$ results in the polynomial decay at infinity of the Fourier transform.

Actually we will need to know $\mathcal{F}^{-1}(e^{-t|x|})$, which is $\frac{1}{(2\pi)^n}\mathcal{F}(e^{-t|x|})(-x)$, so

$$\mathcal{F}^{-1}(e^{-t|\xi|}) = \pi^{-(\frac{n+1}{2})}\Gamma\left(\frac{n+1}{2}\right)\frac{t}{(t^2 + |\xi|^2)^{\frac{n+1}{2}}}.$$

5.   Let $f(x) = |x|^\alpha$. For $\alpha > -n$ (we may even take α complex if Re $\alpha > -n$), f is locally integrable (it is never integrable) and does not increase too fast so it defines a tempered distribution. To compute its Fourier transform we use the same method as in the previous example. We have

$$\int_0^\infty s^{-\frac{\alpha}{2}-1}e^{-s|x|^2}ds = |x|^\alpha\int_0^\infty s^{-\frac{\alpha}{2}-1}e^{-s}ds$$

$$= \Gamma\left(-\frac{\alpha}{2}\right)|x|^\alpha$$

(we have made the change of variable $s \rightarrow s|x|^{-2}$). Of course for this integral to converge, the singularity at $s = 0$ must be better than s^{-1}, so we require $\alpha < 0$. Thus we have imposed the conditions $-n < \alpha < 0$ to obtain

$$|x|^\alpha = \frac{1}{\Gamma(-\frac{\alpha}{2})}\int_0^\infty s^{-\frac{\alpha}{2}-1}e^{-s|x|^2}ds.$$

Now we may compute

$$\mathcal{F}(|x|^\alpha) = \frac{1}{\Gamma\left(-\frac{\alpha}{2}\right)}\int_0^\infty s^{-\frac{\alpha}{2}-1}\mathcal{F}(e^{-s|x|^2})\,ds$$

$$= \frac{\pi^{n/2}}{\Gamma\left(-\frac{\alpha}{2}\right)}\int_0^\infty s^{-\frac{\alpha}{2}-\frac{n}{2}-1}e^{-|\xi|^2/4s}ds.$$

Now to evaluate the integral make the change of variable $s \rightarrow |\xi|^2/4s, ds \rightarrow \frac{|\xi|^2}{4s^2}ds$ so

$$\mathcal{F}(|x|^\alpha) = \frac{\pi^{n/2}}{\Gamma(-\frac{\alpha}{2})}\int_0^\infty \left[\frac{|\xi|^2}{4s}\right]^{-\frac{\alpha}{2}-\frac{n}{2}-1}e^{-s\frac{|\xi|^2}{4s^2}}\frac{|\xi|^2}{4s^2}\,ds$$

$$= \frac{\pi^{n/2}2^{\alpha+n}}{\Gamma(-\frac{\alpha}{2})}|\xi|^{-\alpha-n}\int_0^\infty s^{\frac{\alpha}{2}+\frac{n}{2}-1}ds$$

$$= \frac{\pi^{n/2}2^{\alpha+n}\Gamma(\frac{\alpha}{2}+\frac{n}{2})}{\Gamma(-\frac{\alpha}{2})}|\xi|^{-\alpha-n}.$$

Note that $-\alpha-n$ satisfies the same conditions as α, namely $-n < -n-\alpha < 0$. We mention one special case that we will use later: $n = 3$ and $\alpha = -1$. Here we have

$$
\begin{aligned}
\mathcal{F}(|x|^{-1}) &= \frac{\pi^{3/2}2^2\Gamma(1)}{\Gamma\left(\frac{1}{2}\right)}|\xi|^{-2} \\
&= 4\pi|\xi|^{-2}
\end{aligned}
$$

which we can write as

$$
\mathcal{F}^{-1}(|\xi|^{-2}) = \frac{1}{4\pi|x|}.
$$

4.3 Convolutions with tempered distributions

Many applications of the Fourier transform to solve differential equations lead to convolutions where one factor is a tempered distribution. Recall

$$
\varphi * \psi(x) = \int \varphi(x - y)\psi(y)\,dy
$$

if $\varphi, \psi \in \mathcal{S}$ defines a function in $\mathcal{S}$ and $\mathcal{F}(\varphi * \psi) = \hat{\varphi} \cdot \hat{\psi}$. Since products are not defined for all distributions we cannot expect to define convolutions of two tempered distributions. However if one factor is in $\mathcal{S}$ there is no problem. Fix $\psi \in \mathcal{S}$. Then convolution with ψ is an operation that preserves $\mathcal{S}$, so to define $\psi * f$ for $f \in \mathcal{S}'$ we need only find an adjoint identity. Now

$$
\int \psi * \varphi_1(x)\varphi_2(x)\,dx = \iint \psi(x - y)\varphi_1(y)\varphi_2(x)\,dy\,dx.
$$

If we do the x-integration first,

$$
\int \psi(x - y)\varphi_2(x)\,dx = \tilde{\psi} * \varphi_2(y)
$$

where $\tilde{\psi}(x) = \psi(-x)$. Thus

$$
\int \psi * \varphi_1(x)\varphi_2(x)\,dx = \int \varphi_1(y)\tilde{\psi} * \varphi_2(y)\,dy,
$$

which is our adjoint identity. Thus we define $\psi * f$ by

$$
\langle \psi * f, \varphi \rangle = \langle f, \tilde{\psi} * \varphi \rangle.
$$

From this we obtain $\mathcal{F}(\psi * f) = \hat{\psi} \cdot \hat{f}$ by definition chasing:

$$
\langle \mathcal{F}(\psi * f), \varphi \rangle = \langle \psi * f, \hat{\varphi} \rangle = \langle f, \tilde{\psi} * \hat{\varphi} \rangle = \langle \hat{f}, \mathcal{F}^{-1}(\tilde{\psi} * \hat{\varphi}) \rangle.
$$

Now $\mathcal{F}^{-1}(\tilde{\psi} * \hat{\varphi}) = ((2\pi)^n \mathcal{F}^{-1}\tilde{\psi}) \cdot \varphi$ and $(2\pi)^n \mathcal{F}^{-1}\tilde{\psi} = \hat{\psi}$ so $\langle \mathcal{F}(\psi * f), \varphi \rangle = \langle \hat{f}, \hat{\psi} \cdot \varphi \rangle = \langle \hat{\psi} \cdot \hat{f}, \varphi \rangle$, which shows $\mathcal{F}(\psi * f) = \hat{\psi} \cdot \hat{f}$.

There is another way to define the convolution, however, which is much more direct. Remember that if $f \in \mathcal{S}$ then

$$\psi * f(x) = \int \psi(x - y) f(y) \, dy.$$

It is suggestive to write this

$$\psi * f(x) = \langle f, \tau_{-x} \tilde{\psi} \rangle$$

where $\tau_{-x} \tilde{\psi}(y) = \psi(x - y)$ is still in $\mathcal{S}$. But written this way it makes sense for any tempered distribution f. Of course this defines $\psi * f$ as a function, in fact a C^∞ function, since we can put all derivatives on ψ. What ought to be true is that the distribution defined by this function is tempered and agrees with the previous definition. This is in fact the case. What you have to show is that if we denote by $g(x) = \langle f, \tau_{-x} \tilde{\psi} \rangle$ then $\int g(x) \varphi(x) \, dx = \langle f, \tilde{\psi} * \varphi \rangle$. Formally we can derive this by substituting

$$
\begin{aligned}
\int g(x) \varphi(x) \, dx &= \int \langle f, \tau_{-x} \tilde{\psi} \rangle \varphi(x) \, dx \\
&= \left\langle f, \int (\tau_{-x} \tilde{\psi}) \varphi(x) \, dx \right\rangle
\end{aligned}
$$

and then noting that

$$\int (\tau_{-x} \tilde{\psi}(y)) \varphi(x) \, dx = \int \tilde{\psi}(y - x) \varphi(x) \, dx = \tilde{\psi} * \varphi(y).$$

What this shows is that convolution is a smoothing process. If you start with any tempered distribution, no matter how rough, and take the convolution with a test function, you get a smooth function.

Let us look at some simple examples. If $f = \delta$ then

$$\psi * \delta(x) = \langle \delta, \tau_{-x} \tilde{\psi} \rangle = \psi(x - y)|_{y=0} = \psi(x)$$

so $\psi * \delta = \psi$. This is consistent with $\mathcal{F}(\psi * \delta) = \hat{\psi} \cdot \hat{\delta} = \hat{\psi}$ since $\hat{\delta} = 1$. If we differentiate this result we get

$$
\begin{aligned}
\frac{\partial}{\partial x_k} \psi(x) &= \frac{\partial}{\partial x_k} (\psi * \delta) \\
&= \psi * \frac{\partial}{\partial x_k} \delta
\end{aligned}
$$

(the derivative of a convolution with a distribution can be computed by putting the derivative on either factor). This may also be computed directly. What it shows is that differentiation is a special case of convolution—you convolve with a derivative of the δ-function.

We can also reinterpret the Fourier inversion formula as a convolution equation. If we write the double integral for $\mathcal{F}^{-1}\mathcal{F}f$ as an iterated integral in the reverse order,

$$\mathcal{F}^{-1}\mathcal{F}f(x) = \int_{-\infty}^{\infty} f(y)\left(\frac{1}{2\pi}\int_{-\infty}^{\infty} e^{-i(x-y)\cdot\xi}d\xi\right)dy,$$

then it is just the convolution of f with $\frac{1}{2\pi}\int_{-\infty}^{\infty} e^{-ix\cdot\xi}d\xi$ which is the inverse Fourier transform of the constant function 1. But we recognize from $\hat{\delta} \equiv 1$ that $\frac{1}{2\pi}\int_{-\infty}^{\infty} e^{-ix\cdot\xi}d\xi = \delta(x)$ (in the distribution sense, of course) so that

$$\mathcal{F}^{-1}\mathcal{F}f = f * \delta = f.$$

In a sense, the identity

$$\frac{1}{2\pi}\int_{-\infty}^{\infty} e^{-ix\cdot\xi}d\xi = \delta(x)$$

is the Fourier inversion formula for the distribution δ, and this single inversion formula implies the inversion formula for all tempered distributions. We will encounter a similar phenomenon in the next chapter: to solve a differential equation $Pu = f$ for any f it suffices (in many cases) to solve it for $f = \delta$. In this sense δ is the first and best distribution!

4.4   Problems

1. Let

$$f(x) = \begin{cases} e^{-tx} & x > 0 \\ 0 & x \le 0. \end{cases}$$

   Compute $\hat{f}(\xi)$.

2. Let

$$f(x) = x_1|x|^\alpha \text{ in } \mathbb{R}^n \text{ for } \alpha > -n - 1.$$

   Compute $\hat{f}(\xi)$. (*Hint*: Compute $(d/dx_1)|x|^{\alpha+2}$.)

3. Let $f(x) = (1 + |x|^2)^{-\alpha}$ in $\mathbb{R}^n$ for $\alpha > 0$. Show that

$$f(x) = c_\alpha \int_0^{\infty} t^{\alpha-1}e^{-t}e^{-t|x|^2} dt$$

where c_α is a positive constant. Conclude that $\hat{f}(\xi)$ is a positive function.

4. Express the integral

$$F(x) = \int_{-\infty}^{x} f(t)\, dt \text{ for } f \in \mathcal{S}$$

as the convolution of f with a tempered distribution.

5. Let $f(x)$ be a continuous function on $\mathbb{R}^1$ periodic of period 2π. Show that $\hat{f}(\xi) = \sum_{n=-\infty}^{\infty} b_n \tau_n \delta$ and relate b_n to the coefficients of the Fourier series of f.

6. What is the Fourier transform of x^k on $\mathbb{R}^1$?

7. Show that $\mathcal{F}(d_r f) = r^{-n} d_{1/r} \mathcal{F} f$ for tempered distributions (cf. problem 3.6.2).

8. Show that if f is homogeneous of degree t then $\mathcal{F} f$ is homogeneous of degree $-n - t$.

9. Let

$$\langle f, \varphi \rangle = \lim_{\epsilon \to 0} \int_{|x| > \epsilon} \frac{\varphi(x)}{x}\, dx \text{ in } \mathbb{R}^1.$$

Show that $\mathcal{F} f(\xi) = c \operatorname{sgn} \xi$ because $\mathcal{F} f$ is odd and homogeneous of degree zero. Compute the constant c by using $d/d\xi \operatorname{sgn} \xi = 2\delta$. (Convolution with f is called the "Hilbert transform".)

10. Compute the Fourier transform of $\operatorname{sgn} x\, e^{-t|x|}$ on $\mathbb{R}^1$. Take the limit as $t \to 0$ to compute the Fourier transform of $\operatorname{sgn} x$. Compare the result with problem 9.

11. Use the Fourier inversion formula to "evaluate" $\int_{-\infty}^{\infty} \frac{\sin x}{x}\, dx$ (this integral converges as an improper integral

$$\lim_{N \to \infty} \int_{-N}^{N} \frac{\sin x}{x}\, dx$$

to the indicated value, although this formal computation is not a proof). (*Hint*: See problem 3.6.6.) Use the Plancherel formula to evaluate

$$\int_{-\infty}^{\infty} \left(\frac{\sin x}{x} \right)^2 dx.$$

12. Use the Plancherel formula to evaluate

$$\int_{-\infty}^{\infty} \frac{1}{(1+x^2)^2}\, dx.$$

13. Compute the Fourier transform of $1/(1+x^2)^2$ in $\mathbb{R}^1$.

14. Compute the Fourier transform of

$$f(x) = \begin{cases} x^k e^{-x} & x > 0 \\ 0 & x \le 0. \end{cases}$$

Can you do this even when k is not an integer (but $k > 0$)?

Chapter 5

Solving Partial Differential Equations

5.1 The Laplace equation

Recall that Δ stands for

$$\frac{\partial^2}{\partial x_1^2} + \frac{\partial^2}{\partial x_2^2}$$

in $\mathbb{R}^2$ or

$$\frac{\partial^2}{\partial x_1^2} + \frac{\partial^2}{\partial x_2^2} + \frac{\partial^2}{\partial x_3^2}$$

in $\mathbb{R}^3$. First we ask if there are solutions to the equation $\Delta u = f$ for a given f. If there are, they are not unique, for we can always add a harmonic function (solution of $\Delta u = 0$) without changing the right-hand side.

Now suppose we could solve the equation $\Delta P = \delta$. Then

$$\Delta(P * f) = \Delta P * f = \delta * f = f$$

so $P * f$ is a solution of $\Delta u = f$. Of course we have reasoned only formally, but if P turns out to be a tempered distribution and $f \in \mathcal{S}$, then every step is justified.

Such solutions P are called *fundamental solutions* or *potentials* and have been known for centuries. We have already found a potential when $n = 2$. Remember we found $\Delta \log(x_1^2 + x_2^2) = \delta/4\pi$ so that $\log(x_1^2 + x_2^2)/4\pi$ is a potential (called the *logarithmic potential*).

When $n = 3$ we can solve $\Delta P = \delta$ by taking the Fourier transform of both sides. We get

$$\mathcal{F}(\Delta P) = -|\xi|^2 \hat{P}(\xi) = \mathcal{F}\delta = 1.$$

So $\hat{P}(\xi) = -|\xi|^{-2}$ and we have computed (example 5 of section 4.2)

$$P(x) = -\mathcal{F}^{-1}(|\xi|^{-2}) = -\frac{1}{4\pi|x|}$$

(this is called the *Newtonian potential*). We could also verify directly using Stokes' theorem that $\Delta P = \delta$ in this case.

Now there is one point that should be bothering you about the above computation. We said that the solution was not unique, and yet we came up with just one solution. There are two explanations for this. First, we did cheat a little. From the equation $-|\xi|^2\hat{P} = 1$ we cannot conclude that $\hat{P} = -1/|\xi|^2$ because the multiplication is not of two functions but a function times a distribution. Now it is true that $-|\xi|^2 \cdot (-|\xi|^{-2}) = 1$ regarding $-|\xi|^{-2}$ as a distribution. But if we write $\hat{P} = -|\xi|^{-2} + g$ then $-|\xi|^2\hat{P} = 1$ is equivalent to $-|\xi|^2 g = 0$ and this equation has nonzero solutions. For instance, $g = \delta$ is a solution since

$$\langle -|\xi|^2\delta, \varphi \rangle = \langle \delta, -|\xi|^2\varphi \rangle = -|\xi|^2\varphi(\xi)|_{\xi=0} = 0 \cdot \varphi(0) = 0.$$

This leads to the fundamental solution

$$-\frac{1}{4\pi|x|} + \frac{1}{(2\pi)^3}.$$

More generally we are allowed to take all possible solutions of $-|\xi|^2 g = 0$. It is apparent that such distributions must be concentrated at $\xi = 0$, and later we will show they all are finite linear combinations of derivatives of the δ-function (note however that only some distributions of this form satisfy $-|\xi|^2 g = 0$; $g = (\partial^2/\partial x_1^2)\delta$ does not). Taking $\mathcal{F}^{-1}(-|\xi|^{-2} + g)$ we obtain the Newtonian potential plus a polynomial that is harmonic.

This is still not the whole story, for we know that the general solution of $\Delta u = \delta$ is the Newtonian potential plus a harmonic function. There are many harmonic functions that are not polynomials. So how have these solutions escaped us? To put the paradox more starkly, if we attempt to describe all harmonic functions on $\mathbb{R}^3$ (the same argument works for $\mathbb{R}^2$ as well) by using the Fourier transforms to solve $\Delta u = 0$, we obtain $-|\xi|^2\hat{u}(\xi) = 0$, from which we deduce that u must be a polynomial. That seems to exclude functions that are harmonic but are not polynomials, such as $e^{x_1}\cos x_2$.

But there is really no contradiction because $e^{x_1}\cos x_2$ is not a tempered distribution; it grows too fast as $x_1 \to \infty$, so its Fourier transform is not defined. In fact what we have shown is that *any* harmonic function that is not a polynomial must grow too fast at infinity to be a tempered distribution. Stating this in the contrapositive form, if a harmonic function on $\mathbb{R}^2$

or $\mathbb{R}^3$ is of polynomial growth

$$(|u(x)| \leq c|x|^N \text{ as } x \to \infty)$$

then it must be a polynomial. This is a generalization of Liouville's theorem (a bounded entire analytic function is constant).

The two points I have just made bear repeating in a more general context: (1) when solving an equation for a distribution by division, there will be extra solutions at the zeroes of the denominator, and (2) when using Fourier transforms to solve differential equations you will only obtain those solutions that do not grow too rapidly at infinity.

Now we return to the Laplace equation. We have seen that $\Delta(P*f) = f$ if $f \in \mathcal{S}$. Actually this solution is valid for more general functions f, as long as the convolution can be reasonably defined. For instance, since P is locally integrable in both cases, $P * f(x) = \int P(x-y)f(y)\,dy$ makes sense for f continuous and vanishing outside a bounded set, and $\Delta(P*f) = f$.

Usually one is interested in finding not just a solution to a differential equation, but a solution that satisfies certain side conditions which determine it uniquely. A typical example is the following: Let D be a bounded domain in $\mathbb{R}^2$ or $\mathbb{R}^3$ with a smooth boundary B (in $\mathbb{R}^2$ B is a curve, in $\mathbb{R}^3$ B is a surface). Let f be a continuous function on D (continuous up to the boundary) and g a continuous function on B. We then seek solution of $\Delta u = f$ in D with $u = g$ on B.

To solve this problem first extend f so that it is defined outside of D. The simplest way to do this is to set it equal to zero outside D—this results in a discontinuous function, but that turns out not to matter. Call the extension F and look at $P * F$. Since $\Delta(P * F) = F$ and $F = f$ on D we set $v = P * F$ restricted to D and so $\Delta v = f$ on D. Calling $w = u - v$ we see that w must satisfy

$$\Delta w = 0 \text{ on } D$$

$$w = g - h \text{ on } B$$

where $h = P * F$ restricted to B. Now it can be shown that h is continuous so that the problem for w is the classical Dirichlet problem: find a harmonic function on D with prescribed continuous values on B. This problem always has a unique solution, and for some domains D it is given by explicit integrals. Once you have the unique solution w to the Dirichlet problem, $u = v + w$ is the unique solution to the original problem.

Next we will use Fourier transforms to study the Dirichlet problem when D is a half-plane (D is not bounded, of course, but it is the easiest domain to study). It will be convenient to change notation here. We let t be a real variable that will always be ≥ 0, and we let $x = (x_1, \ldots, x_n)$ be a variable in $\mathbb{R}^n$ (the cases of physical interest are $n = 1, 2$). We consider functions $u(x,t)$ for $x \in \mathbb{R}^n, t \geq 0$ which are harmonic

$$\left[\frac{\partial^2}{\partial t^2} + \frac{\partial^2}{\partial x_1^2} + \cdots + \frac{\partial^2}{\partial x_n^2} \right] u = \left[\frac{\partial^2}{\partial t^2} + \Delta_x \right] u = 0$$

and which take prescribed values on the boundary $t = 0 : u(x,0) = f(x)$. For now let us take $f \in \mathcal{S}(\mathbb{R}^n)$.

The solution is not unique for we may always add ct, that is a harmonic function that vanishes on the boundary. To get a unique solution we must add a growth condition at infinity, say that u is bounded.

The method we use is to take Fourier transforms in the x-variables only (this is sometimes called the partial Fourier transform). That is, for each fixed $t \geq 0$, we regard $u(x,t)$ as a function of x. Since it is bounded it defines a tempered distribution and so it has a Fourier transform that we denote $\mathcal{F}_x u(\xi, t)$ (sometimes the more ambiguous notation $\hat{u}(\xi, t)$ is used). The differential equation

$$\frac{\partial^2}{\partial t^2} u(x,t) + \Delta_x u(x,t) = 0$$

becomes

$$\frac{\partial^2}{\partial t^2} \mathcal{F}_x u(\xi, t) - |\xi|^2 \mathcal{F}_x u(\xi, t) = 0$$

and the boundary condition $u(x,0) = f(x)$ becomes $\mathcal{F}_x u(\xi, 0) = \hat{f}(\xi)$.

Now what have we gained by this? We have replaced a partial differential equation by an ordinary differential equation, since only t-derivatives are involved. And the ordinary differential equation is so simple it can be solved directly. For each fixed ξ (this is something of a cheat, since $\mathcal{F}_x u(\xi, t)$ is only a distribution, not a function of ξ; however, in this case we get the right answer in the end), the equation

$$\frac{\partial^2}{\partial t^2} \mathcal{F}_x u(\xi, t) - |\xi|^2 \mathcal{F}_x u(\xi, t) = 0$$

has solutions $c_1 e^{t|\xi|} + c_2 e^{-t|\xi|}$ where c_1 and c_2 are constants. Since these constants can change with ξ we should write

$$\mathcal{F}_x u(\xi, t) = c_1(\xi) e^{t|\xi|} + c_2(\xi) e^{-t|\xi|}$$

for the general solution. Now we can simplify this formula by considering the fact that we want $u(x,t)$ to be bounded. The term $c_1(\xi) e^{t|\xi|}$ is going to grow with t as $t \to \infty$, unless $c_1(\xi) = 0$. So we are left with

$\mathcal{F}_x u(\xi, t) = c_2(\xi)e^{-t|\xi|}$. From the boundary condition $\mathcal{F}_x u(\xi, 0) = \hat{f}(\xi)$ we obtain $c_2(\xi) = \hat{f}(\xi)$ so $\mathcal{F}_x u(\xi, t) = \hat{f}(\xi)e^{-t|\xi|}$, hence

$$u(x, t) = \mathcal{F}_x^{-1}(\hat{f}(\xi)e^{-t|\xi|}) = \mathcal{F}_x^{-1}(e^{-t|\xi|}) * f(x).$$

Now in example (4) of 4.2 we computed

$$\mathcal{F}^{-1}(e^{-t|\xi|}) = \pi^{-(\frac{n+1}{2})}\Gamma\left(\frac{n+1}{2}\right)\frac{t}{(t^2 + |x|^2)^{\frac{n+1}{2}}}$$

so

$$u(x, t) = \pi^{-(\frac{n+1}{2})}\Gamma\left(\frac{n+1}{2}\right)\int_{\mathbb{R}^n} f(y)\frac{t}{(t^2 + (x-y)^2)^{\frac{n+1}{2}}}\, dy.$$

This is referred to as the Poisson integral formula for the half-space. The integral is convergent as long as f is a bounded function and gives a bounded harmonic function with boundary values f. The derivation we gave involved some questionable steps; however, the validity of the result can be checked and the uniqueness proved by other methods.

Exercise: Verify that u is harmonic by differentiating the integral. (Note that the denominator is never zero since $t > 0$.)
  The special case $n = 1$

$$u(x, t) = \frac{1}{\pi}\int_{-\infty}^{\infty} f(y)\frac{t}{t^2 + (x-y)^2}\, dy$$

can be derived from the Poisson integral formula for the disk by conformal mapping.

5.2   The heat equation

We retain the notation from the previous case: $t \geq 0, x \in \mathbb{R}^n, u(x, t)$. In this case the heat equation is

$$\frac{\partial}{\partial t}u(x, t) = k\Delta_x u(x, t)$$

where k is a positive constant. You should think of t as time, $t = 0$ the initial time, x a point in space ($n = 1, 2, 3$ are the physically interesting cases), and $u(x, t)$ temperature. The boundary condition $u(x, 0) = f(x), f$ given in $\mathcal{S}$, should be thought of as the initial temperature. From physical reasoning there should be a unique solution. Actually for uniqueness we need some additional growth condition on the solution—boundedness is more than adequate (although it requires some work to exhibit a nonzero solution with zero initial conditions).

We can find the solution explicitly by the method of partial Fourier transform. The differential equation becomes

$$\frac{\partial}{\partial t}\mathcal{F}_x u(\xi, t) = -k|\xi|^2 \mathcal{F}_x u(\xi, t)$$

and the initial condition becomes $\mathcal{F}_x(\xi, 0) = \hat{f}(\xi)$. Solving the differential equation we have

$$\mathcal{F}_x u(\xi, t) = c(\xi) e^{-kt|\xi|^2}$$

and from the initial condition $c(\xi) = \hat{f}(\xi)$ so that

$$
\begin{aligned}
u(x, t) &= \mathcal{F}^{-1}(e^{-kt|\xi|^2} \hat{f}(\xi)) \\
&= \mathcal{F}^{-1}(e^{-kt|\xi|^2}) * f \\
&= \frac{1}{(4\pi kt)^{n/2}} \int e^{-|x-y|^2/4kt} f(y) \, dy.
\end{aligned}
$$

Exercise: Verify directly that this gives a solution of the heat equation.

One interesting aspect of this solution is the way it behaves with respect to time. This is easiest to see on the Fourier transform side:

$$\mathcal{F}_x u(\xi, t) = e^{-kt|\xi|^2} \hat{f}(\xi)$$

decreases at infinity more rapidly as t increases. This decrease at infinity corresponds roughly to "smoothness" of $u(x, t)$. Thus as time increases, so does the smoothness of the temperature. The other side of the coin is that if we try to reverse time we run into trouble. In other words, if we try to find the solution for negative t (corresponding to times before the initial measurement), the initial temperature $f(x)$ must be very smooth (so that $\hat{f}(\xi)$ decreases so fast that $\hat{f}(\xi)e^{-kt|\xi|^2}$ is a tempered distribution). Even if the solution does exist for negative t, it is not given by a simple formula (the formula we derived is definitely nonsense for $t < 0$).

So far, the problems we have looked at involve solving a differential equation on an unbounded region. Most physical problems involve bounded regions. For the heat equation, the simplest physically realistic domain is to take $n = 1$ and let x vary in a finite interval, so $0 \le x \le 1$. This requires that we formulate some sort of boundary conditions at $x = 0$ and $x = 1$. We will take *periodic* boundary conditions

$$u(0, t) = u(1, t) \text{ all } t$$

which correspond to a circular piece of wire (insulated, so heat does not transfer to the ambient space around the wire). In the problems you will encounter other boundary conditions.

The word "periodic" is the key to the solution. We imagine the initial temperature $f(x)$, which is defined on $[0, 1]$ (to be consistent with the periodic boundary conditions we must have $f(0) = f(1)$), extended to the whole line as a periodic function of x, $f(x + 1) = f(x)$ all x, and similarly for $u(x, t)$, $u(x + 1, t) = u(x, t)$ all x (note that the periodic condition on u implies the boundary condition just by setting $x = 0$).

Now if we substitute our periodic function f in the solution for the whole line, we obtain

$$
\begin{aligned}
u(x,t) &= \frac{1}{(4\pi kt)^{1/2}} \int_{-\infty}^{\infty} e^{-(x-y)^2/4kt} f(y)\, dy \\
&= \sum_{j=-\infty}^{\infty} \frac{1}{(4\pi kt)^{1/2}} \int_{0}^{1} e^{-(x+j-y)^2/4kt} f(y)\, dy
\end{aligned}
$$

(break the line up into intervals $j \le y \le j + 1$ and make the change of variable $y \to y - j$ in each interval, using the periodicity of the f to replace $f(y - j)$ by $f(y)$). We can write this

$$
u(x,t) = \int_{0}^{1} \left(\frac{1}{(4\pi kt)^{1/2}} \sum_{j=-\infty}^{\infty} e^{-(x+j-y)^2/4kt} \right) f(y)\, dy
$$

because the series converges rapidly. Observe that this formula does indeed produce a periodic function $u(x, t)$, since the substitution $x \to x + 1$ can be erased by the change of summation variable $j \to j - 1$.

Perhaps you are more familiar with a solution to the same problem using Fourier series. This, in fact, was one of the first problems that led Fourier to the discovery of Fourier series. Since the problem has a unique solution, the two solutions must be equal, but they are not the same. The Fourier series solution looks like

$$
\begin{aligned}
u(x,t) &= \sum_{n=-\infty}^{\infty} a_n e^{-4\pi^2 n^2 kt} e^{2\pi inx} \\
a_n &= \int_{0}^{1} f(x) e^{-2\pi inx}\, dx.
\end{aligned}
$$

Both solutions involve infinite series, but the one we derived has the advantage that the terms are all positive (if f is positive).

5.3   The wave equation

The equation is $(\partial^2/\partial t^2)u(x,t) = k^2\Delta_x u(x,t)$ where t is now any real number and $x \in \mathbb{R}^n$. We will consider the cases $n = 1,2,3$ only, which describe roughly vibrations of a string, a drum, and sound waves in air, respectively. The constant k is the maximum propagation speed, as we shall see shortly. The initial conditions we give are for both u and $\partial u/\partial t$,

$$u(x,0) = f(x), \qquad \frac{\partial u}{\partial t}(x,0) = g(x).$$

As usual we take f and g in $\mathcal{S}$, although the solution we obtain allows much more general choice. These initial conditions (called *Cauchy data*) determine a unique solution without any growth conditions.

We solve by taking partial Fourier transforms. We obtain

$$\frac{\partial^2}{\partial t^2}\mathcal{F}_x u(\xi,t) = -k^2|\xi|^2\mathcal{F}_x u(\xi,t)$$

$$\mathcal{F}_x u(\xi,0) = \hat{f}(\xi) \qquad \frac{\partial}{\partial t}\mathcal{F}_x u(\xi,0) = \hat{g}(\xi).$$

The general solution of the differential equation is

$$c_1(\xi)\cos kt|\xi| + c_2(\xi)\sin kt|\xi|$$

and from the initial conditions we obtain

$$\mathcal{F}_x u(\xi,t) = \hat{f}(\xi)\cos kt|\xi| + \hat{g}(\xi)\frac{\sin kt|\xi|}{k|\xi|}.$$

Before inverting the Fourier transform let us make some observations about the solution. First, it is clear that time is reversible—except for a minus sign in the second term, there is no difference between t and $-t$. So the past is determined by the present as well as the future.

Another thing we can see, although it requires work, is the conservation of energy. The energy of the solution $u(x,t)$ at time t is defined as

$$E(t) = \frac{1}{2}\int_{\mathbb{R}^n}\left(\left|\frac{\partial}{\partial t}u(x,t)\right|^2 + k^2\sum_{j=1}^{n}\left|\frac{\partial}{\partial x_j}u(x,t)\right|^2\right)dx.$$

The first term is kinetic energy and the second is potential energy. Conservation of energy says that $E(t)$ is independent of t.

To see this we express the energy in terms of $\mathcal{F}_x u(\xi,t)$. Since

$$\mathcal{F}_x\left(\frac{\partial}{\partial x_j}u\right)(\xi,t) = -i\xi_j\mathcal{F}_x u(\xi,t)$$

we have

$$E(t) = \frac{1}{2(2\pi)^n} \int_{\mathbb{R}^n} \left(\left| \frac{\partial}{\partial t} \mathcal{F}_x u(\xi, t) \right|^2 + k^2 |\xi|^2 |\mathcal{F}_x u(\xi, t)|^2 \right) d\xi$$

by the Plancherel formula. Now

$$\begin{aligned}
|\mathcal{F}_x u(\xi, t)|^2 &= \left(\hat{f}(\xi) \cos kt|\xi| + \hat{g}(\xi) \frac{\sin kt|\xi|}{k|\xi|} \right) \\
&\cdot \left(\overline{\hat{f}(\xi)} \cos kt|\xi| + \overline{\hat{g}(\xi)} \frac{\sin kt|\xi|}{k|\xi|} \right)
\end{aligned}$$

and

$$\begin{aligned}
\left| \frac{\partial}{\partial t} \mathcal{F}_x u(\xi, t) \right|^2 &= (-k|\xi| \hat{f}(\xi) \sin kt|\xi| + \hat{g}(\xi) \cos kt|\xi|) \\
&\cdot (-k|\xi| \overline{\hat{f}(\xi)} \sin kt|\xi| + \overline{\hat{g}(\xi)} \cos kt(\xi))
\end{aligned}$$

so that

$$\left| \frac{\partial}{\partial t} \mathcal{F}_x u(\xi, t) \right|^2 + k^2 |\xi|^2 |\mathcal{F}_x u(\xi, t)|^2 = k^2 |\xi|^2 |\hat{f}(\xi)|^2 + |\hat{g}(\xi)|^2$$

(the cross terms cancel and the $\sin^2 + \cos^2$ terms add to one). Thus

$$\begin{aligned}
E(t) &= \frac{1}{2(2\pi)^n} \int (k^2 |\xi|^2 |\hat{f}(\xi)|^2 + |\hat{g}(\xi)|^2) \, d\xi \\
&= \frac{1}{2} \int \left(k^2 \sum_{j=1}^{n} \left| \frac{\partial}{\partial x_j} f(x) \right|^2 + |g(x)|^2 \right) dx
\end{aligned}$$

independent of t.

Now to invert the Fourier transform. When $n = 1$ this is easy since $\cos kt|\xi| = \frac{1}{2}(e^{ikt\xi} + e^{-ikt\xi})$ so

$$\mathcal{F}^{-1}(\cos kt|\xi| \hat{f}(\xi))(x) = \frac{1}{2}(f(x + kt) + f(x - kt)).$$

Similarly

$$\mathcal{F}^{-1}\left(\hat{g}(\xi) \frac{\sin kt|\xi|}{k|\xi|} \right) = \frac{1}{2k} \int_{-kt}^{kt} g(x + s) \, ds.$$

When $n = 3$ the answer is given in terms of surface integrals over spheres. Let σ denote the distribution

$$\langle \sigma, \varphi \rangle = \int_{|x|=1} \varphi(x) \, d\sigma(x)$$

where $d\sigma(x)$ is the element of surface integration on the unit sphere. In terms of spherical coordinates $(x, y, z) = (\cos\theta_1, \sin\theta_1 \cos\theta_2, \sin\theta_1 \sin\theta_2)$ for the sphere, with $0 \le \theta_1 \le \pi$ and $0 \le \theta_2 \le 2\pi$, this is just

$$\langle \sigma, \varphi \rangle = \int_0^{2\pi} \int_0^\pi \varphi(\cos\theta_1, \sin\theta_1 \cos\theta_2, \sin\theta_1 \sin\theta_2) \sin\theta_1 d\theta_1, d\theta_2.$$

Now to compute $\hat{\sigma}(\xi)$ we need to evaluate this integral when $\varphi(x) = e^{ix\cdot\xi}$. To make the computation easier we use the observation (from problem 3.8) that $\hat{\sigma}$ is radial, so it suffices to compute $\hat{\sigma}(\xi_1, 0, 0)$, for then $\tilde{\sigma}(\xi_1, \xi_2, \xi_3) = \hat{\sigma}(|\xi|, 0, 0)$. But

$$\begin{aligned}
\langle \sigma, e^{ix_1\xi_1} \rangle &= \int_0^{2\pi} \int_0^\pi e^{i\xi_1 \cos\theta_1} \sin\theta_1 d\theta_1 d\theta_2 \\
&= 2\pi \int_0^\pi e^{i\xi_1 \cos\theta_1} \sin\theta_1 d\theta_1 \\
&= \frac{-2\pi e^{i\xi_1 \cos\theta_1}}{i\xi_1} \Big|_0^\pi \\
&= \frac{4\pi \sin\xi_1}{\xi_1}
\end{aligned}$$

and so $\hat{\sigma}(\xi) = 4\pi \sin|\xi|/|\xi|$. Similarly, if σ_r denotes the surface integral over the sphere $|x| = r$ of radius r, then $\hat{\sigma}_r(\xi) = 4\pi r \sin r|\xi|/|\xi|$ and so

$$\mathcal{F}\left(\frac{1}{4\pi k^2 t} \sigma_{kt} \right) = \frac{\sin kt|\xi|}{k|\xi|}.$$

Thus

$$\mathcal{F}^{-1}\left(\hat{g}(\xi) \frac{\sin kt|\xi|}{k|\xi|} \right) = \frac{1}{4\pi k^2 t} \sigma_{kt} * g(x).$$

Furthermore, if we differentiate this identity with respect to t we find

$$\mathcal{F}^{-1}(\hat{f}(\xi) \cos kt(|\xi|)) = \frac{\partial}{\partial t} \left(\frac{1}{4\pi k^2 t} \sigma_{kt} * f(x) \right)$$

(we renamed the function f). Thus the solution to the wave equation in $n = 3$ dimensions is simply

$$u(x, t) = \frac{\partial}{\partial t} \left(\frac{1}{4\pi k^2 t} \sigma_{kt} * f(x) \right) + \frac{1}{4\pi k^2 t} \sigma_{kt} * g(x).$$

The convolution can be written directly as

$$\sigma_{kt} * f(x) = \int_{|y|=kt} f(x + y) \, d\sigma_{kt}(y),$$

or it can be expressed in terms of integration over the unit sphere

$$\sigma_{kt} * f(x) = k^2 t^2 \int_{|y|=1} f(x + kt\,y)\,d\sigma$$

$$= k^2 t^2 \int_0^{2\pi} \int_0^{\pi} f\left(x_1 + kt \cos\theta_1, x_2 + kt \sin\theta_1 \cos\theta_2,\right.$$

$$\left. x_3 + kt \sin\theta_1 \sin\theta_2\right) \sin\theta_1 d\theta_1 d\theta_2.$$

When $n = 2$ the solution to the wave equation is easiest to obtain by the so-called "method of descent". We take our initial position and velocity $f(x_1, x_2)$ and $g(x_1, x_2)$ and pretend they are functions of three variables (x_1, x_2, x_3) that are independent of the third variable x_3. Fair enough. We then solve the 3-dimensional wave equation for this initial data. The solution will also be independent of the third variable and will be the solution of the original 2-dimensional problem. This gives us explicitly

$$u(x, t) =$$

$$\frac{\partial}{\partial t}\left(\frac{t}{4\pi} \int_0^{2\pi} \int_0^{\pi} f(x_1 + kt \cos\theta_1, x_2 + kt \sin\theta_1 \cos\theta_2) \sin\theta_1 \, d\theta_1 \, d\theta_2\right)$$

$$+ \frac{t}{4\pi} \int_0^{2\pi} \int_0^{\pi} g(x_1 + kt \cos\theta_1, x_2 + kt \sin\theta_1 \cos\theta_2) \sin\theta_1 \, d\theta_1 \, d\theta_2.$$

There is another way to express this solution. The pair of variables $(\cos\theta_1, \sin\theta_1 \cos\theta_2)$ describes the unit disk $x_1^2 + x_2^2 \leq 1$ in a two-to-one fashion (two different values of θ_2 give the same value to $\cos\theta_2$) as (θ_1, θ_2) vary over $0 \leq \theta_1 \leq \pi, 0 \leq \theta_2 \leq 2\pi$. Thus if we make the substitution $(y_1, y_2) = (\cos\theta_1, \sin\theta_1 \cos\theta_2)$ then $dy_1 dy_2 = \sin^2\theta_1 |\sin\theta_2| d\theta_1 d\theta_2$ and $\sin\theta_1 |\sin\theta_2| = \sqrt{1 - |y|^2}$ so

$$u(x, t) = \frac{\partial}{\partial t}\left(\frac{t}{2\pi} \int_{|y|\leq 1} \frac{f(x + kty)}{\sqrt{1 - |y|^2}}\,dy\right)$$

$$+ \frac{t}{2\pi} \int_{|y|\leq 1} \frac{g(x + kty)}{\sqrt{1 - |y|^2}}\,dy.$$

Note that these are improper integrals because $(1 - |y|^2)^{-1/2}$ becomes infinite as $|y| \to 1$, but they are absolutely convergent. Still another way to write the integrals is to introduce polar coordinates:

$$u(x, t) = \frac{\partial}{\partial t}\left(\frac{t}{2\pi} \int_0^{2\pi} \int_0^1 f(x_1 + ktr \cos\theta, x_2 + ktr \sin\theta)\frac{r\,dr}{\sqrt{1 - r^2}}\,d\theta\right)$$

$$+ \frac{t}{2\pi} \int_0^{2\pi} \int_0^1 g(x_1 + ktr \cos\theta, x_2 + ktr \sin\theta)\frac{r\,dr}{\sqrt{1 - r^2}}\,d\theta.$$

The convergence of the integral is due to the fact that $\int_0^1 (r\,dr/\sqrt{1-r^2})$ is finite.

There are several astounding qualitative facts that we can deduce from these elegant quantitative formulas. The first is that k is the *maximum speed of propagation of signals*. Suppose we make a "noise" located near a point y at time $t = 0$. Can this noise be "heard" at a point x at a later time t? Certainly not if the distance $(x - y)$ from x to y exceeds kt, for the contribution to $u(x, t)$ from $f(y)$ and $g(y)$ is zero until $kt \geq |x - y|$. This is true in all dimensions, and it is a direct consequence of the fact that $u(x, t)$ is expressed as a sum of convolutions of f and g with distributions that vanish outside the ball of radius kt about the origin. (Compare this with the heat equation, where the "speed of smell" is infinite!) Also, of course, there is nothing special about starting at $t = 0$. The finite speed of sound and light are well-known physical phenomena (light is governed by a system of equations, called *Maxwell's equations*, but each component of the system satisfies the wave equation). But something special happens when $n = 3$ (it also happens when n is odd, $n \geq 3$). After the noise is heard, it moves away and leaves no reverberation (physical reverberations of sound are due to reflections off walls, ground, and objects). This is called *Huyghens' principle* and is due to the fact that distributions we convolve f and g with also vanish *inside* the ball of radius kt. Another way of saying this is that signals propagate at exactly speed k. In particular, if f and g vanish outside a ball of radius R, then after a time $\frac{1}{k}(R + |x|)$, there will be a total silence at point x. This is clearly not the case when $n = 1, 2$ (when $n = 1$ it is true for the initial position f, but not the initial velocity g). This can be thought of as a ripple effect: after the noise reaches point x, smaller ripples continue to be heard. A physical model of this phenomenon is the ripples you see on the surface of a pond, but this is in fact a rather unfair example, since the differential equations that govern the vibrations on the surface of water are nonlinear and therefore quite different from the linear wave equation we have been studying. In particular, the rippling is much more pronounced than it is for the 2-dimensional wave equation.

There is a weak form of Huyghens' principle that does hold in all dimensions: the singularities of the signal propagate at exactly speed k. This shows up in the convolution form of the solution when $n = 2$ in the smoothness of $(1 - |y|^2)^{-1/2}$ everywhere except on the surface of the sphere $|y| = 1$.

Another interesting property is the *focusing of singularities*, which shows up most strikingly when $n = 3$. Since the solution involves averaging over a sphere, we can have relatively mild singularities in the initial data over the whole sphere produce a sharp singularity at the center when they all arrive simultaneously. Assume the initial data is radial: $f(x)$ and $g(x)$ depend only on $|x|$ (we write $f(x) = f(|x|), g(x) = g(|x|)$). Then

$u(0,t) = (\partial/\partial t)(t f(kt)) + t g(kt)$ since

$$\frac{1}{4\pi k^2 t} \int_{|y|=kt} f(y)\, dy = \frac{1}{4\pi k^2 t} 4\pi (kt)^2 f(kt)$$

etc.

It is the appearance of the derivative that can make $u(0,t)$ much worse than f or g. For instance, take $g(x) = 0$ and

$$f(x) = \begin{cases} (1 - |x|)^{1/2} & \text{if } |x| \leq 1 \\ 0 & \text{if } |x| > 1. \end{cases}$$

Then f is continuous, but not differentiable. But $u(0,t) = f(kt) + kt f'(kt)$ tends to infinity at $t = 1/k$, which is the instant when all the singularities are focused.

5.4 Schrödinger's equation and quantum mechanics

The quantum theory of a single particle is described by a complex valued "wave function" $\varphi(x)$ defined on $\mathbb{R}^3$. The only restriction on φ is that $\int_{\mathbb{R}^3} |\varphi(x)|^2 dx$ be finite. Since φ and any constant multiple of φ describe the same physical state, it is convenient, but not necessary, to normalize this integral to be one. The wave function changes with time. If $u(x,t)$ is the wave function at time t then

$$\frac{\partial}{\partial t} u(x,t) = ik\Delta_x u(x,t).$$

This is the free Schrödinger equation. There are additional terms if there is a potential or other physical interaction present. The constant k is related to the mass of the particle and Planck's constant.

The free Schrödinger equation is easily solved with initial condition $u(x,0) = \varphi(x)$. We have $(\partial/\partial t)\mathcal{F}_x u(\xi,t) = -ik|\xi|^2 \mathcal{F}_x u(\xi,t)$ and $\mathcal{F}_x u(\xi,0) = \hat{\varphi}(\xi)$ so that

$$\mathcal{F}_x u(\xi,t) = e^{-ikt|\xi|^2} \hat{\varphi}(\xi).$$

Referring to example (3) of 4.2 we find

$$u(x,t) = \left(\frac{\pi}{kt}\right)^{3/2} e^{\pm\frac{3}{4}\pi i} \int_{\mathbb{R}^3} e^{i|x-y|^2/4kt} \varphi(y)\, dy$$

where the $\pm$ sign is the sign of t (of course the factor $e^{\pm\frac{3}{4}\pi i}$ has no physical significance, by our previous remarks).

Actually the expression for $\mathcal{F}_x u$ is more useful. Notice that $|\mathcal{F}_x u(\xi, t)| = |\hat{\varphi}(\xi)|$ is independent of t. Thus

$$\int_{\mathbb{R}^3} |u(x,t)|^2 \, dx = \frac{1}{(2\pi)^n} \int_{\mathbb{R}^3} |\mathcal{F}_x u(\xi,t)|^2 \, d\xi$$

is independent of t, so once the wave-function is normalized at $t = 0$ it remains normalized.

The interpretation of the wave function is somewhat controversial, but the standard description is as follows: there is an imperfect coupling between physical measurement and the wave function, so that a position measurement of a particle with wave function φ will not always produce the same answer. Instead we have only a probabilistic prediction: the probability that the position vector will measure in a set $A \subseteq \mathbb{R}^3$ is

$$\frac{\int_A |\varphi(x)|^2 \, dx}{\int_{\mathbb{R}^3} |\varphi(x)|^2 \, dx}.$$

We have a similar statement for measurements of momentum. If we choose units appropriately, the probability that the momentum vector will measure in a set $B \subseteq \mathbb{R}^3$ is

$$\frac{\int_B |\hat{\varphi}(x)|^2 \, d\xi}{\int_{\mathbb{R}^3} |\hat{\varphi}(x)|^2 \, d\xi}.$$

Note that $\int_{\mathbb{R}^3} |\hat{\varphi}(\xi)|^2 \, d\xi = (2\pi)^n \int_{\mathbb{R}^3} (\varphi(x))^2 \, dx$ so that the denominator is always finite.

Now what happens as time changes? The position probabilities change in a very complicated way, but $|\mathcal{F}_x u(\xi, t)| = |\hat{\varphi}(\xi)|$ so the momentum probabilities remain the same. This is the quantum mechanical analog of conservation of momentum.

5.5   Problems

1. For the Laplace and the heat equation in the half-space prove via the Plancherel formula that

$$\int_{\mathbb{R}^n} |u(x,t)|^2 \, dx \leq \int_{\mathbb{R}^n} |u(x,0)|^2 \, dx \quad t > 0.$$

   What is the limit of this integral as $t \to 0$ and as $t \to \infty$?

2. For the same equations show $|u(x,t)| \leq \sup_{y \in \mathbb{R}^n} |u(y,0)|$. (*Hint:* Write $u = G_t * f$ and observe that $G_t(x) \geq 0$. Then use the Fourier

inversion formula to compute $\int_{\mathbb{R}^n} G_t(x)\,dx$ and estimate

$$|u(x,t)| \leq \left[\int_{\mathbb{R}^n} G_t(y)\,dy\right] \sup_{y\in\mathbb{R}^n} |f(y)|.$$

3. Solve

$$\left[\frac{\partial^2}{\partial t^2} + \Delta_x\right]^2 u(x,t) = 0$$

for $t \geq 0, x \in \mathbb{R}^n$ given

$$u(x,0) = f(x), \qquad \frac{\partial}{\partial t}u(x,0) = g(x)$$

$f,g \in \mathcal{S}$ with $|u(x,t)| \leq c(1+|t|)$. *Hint:* Show

$$\mathcal{F}_x u(\xi,t) = te^{-t|\xi|}\hat{g}(\xi) + (e^{-t|\xi|} + t|\xi|e^{-t|\xi|})\hat{f}(\xi).$$

To invert note that $\mathcal{F}_x^{-1}(|\xi|e^{-t|\xi|}) = -\frac{\partial}{\partial t}\mathcal{F}_x^{-1}(e^{-t|\xi|})$.

4. Solve $(\partial/\partial t)u(x,t) = k\Delta_x u + u$ for $t \geq 0, x \in \mathbb{R}^n$ with $u(x,0) = f(x) \in \mathcal{S}$.

5. Solve $(\partial^2/\partial t^2)u(x,t) + \Delta_x u(x,t) = 0$ for $0 \leq t \leq T, x \in \mathbb{R}^n$ with $u(x,0) = f(x), u(x,T) = g(x)$ for $\mathcal{F}_x u(\xi,t)$ (do not attempt to invert the Fourier transform).

6. Solve the free Schrödinger equation with initial wave function $\varphi(x) = e^{-|x|^2}$.

7. In two dimensions show that the Laplacian factors

$$\frac{\partial^2}{\partial x^2} + \frac{\partial^2}{\partial y^2} = \left(\frac{\partial}{\partial x} + i\frac{\partial}{\partial y}\right)\left(\frac{\partial}{\partial x} - i\frac{\partial}{\partial y}\right)$$

and the factors commute. Deduce from this that an analytic function is harmonic.

8. Let f be a real-valued function in $\mathcal{S}(\mathbb{R}^1)$ and define g by $\hat{g}(\xi) = -i(\operatorname{sgn}\xi)\hat{f}(\xi)$. Show that g is real-valued and if $u(x,t)$ and $v(x,t)$ are the harmonic functions in the half-plane with boundary values f and g then they are conjugate harmonic functions: $u + iv$ is analytic in $z = x + it$. (*Hint:* Verify the Cauchy-Riemann equations.) Find an expression for v in terms of f. (*Hint:* Evaluate $\mathcal{F}^{-1}(-i\operatorname{sgn}\xi e^{-t|\xi|})$ directly.)

9. Show that a solution to the heat equation (or wave equation) that is independent of time (*stationary*) is a harmonic function of the space variables.

10. Solve the initial value problem for the Klein-Gordon equation

$$\frac{\partial^2 u}{\partial t^2} = \Delta_x u - m^2 u \quad m > 0$$

$$u(x,0) = f(x), \qquad \frac{\partial u}{\partial t}(x,0) = g(x)$$

for $\mathcal{F}_x u(\xi, t)$ (do not attempt to invert the Fourier transform).

11. Show that the energy

$$E(t) = \frac{1}{2} \int_{\mathbb{R}^n} \left(m^2 |u(x,t)|^2 + \left| \frac{\partial}{\partial t} u(x,t) \right|^2 + \sum_{j=1}^{n} \left| \frac{\partial}{\partial x_j} u(x,t) \right|^2 \right) dx$$

is conserved for solutions of the Klein-Gordon equation Klein-Gordon equation.

12. Solve the heat equation on the interval $[0,1]$ with Dirichlet boundary conditions

$$u(0,t) = 0, \quad u(1,t) = 0$$

(*Hint*: Extend all functions by odd reflection about the boundary points $x = 0$ and $x = 1$ and periodicity with period 2).

13. Do the same as problem 12 for Neumann boundary conditions

$$\frac{\partial}{\partial x} u(0,t) = 0, \qquad \frac{\partial}{\partial x} u(1,t) = 0.$$

(*Hint*: This time use even reflections).

14. Show that the inhomogeneous heat equation with homogeneous initial conditions

$$\begin{aligned} \frac{\partial}{\partial t} u(x,t) &= k\Delta_x u(x,t) + F(x,t) \\ u(x,0) &= 0 \end{aligned}$$

is solved by Duhamel's integral

$$u(x,t) = \int_0^t \left(\int_{\mathbb{R}^n} G_s(y) F(x - y, t - s) \, dy \right) ds$$

where

$$G_s(y) = \frac{1}{(4\pi ks)^{n/2}} e^{-|y|^2/4ks}$$

is the solution kernel for the homogeneous heat equation. Use this to solve the fully inhomogeneous problem.

$$\frac{\partial}{\partial t} u(x,t) = k\Delta_x u(x,t) + F(x,t)$$
$$u(x,0) = f(x).$$

15. Show that the inhomogeneous wave equation on $\mathbb{R}^n$ with homogeneous initial data

$$\frac{\partial^2}{\partial t^2} u(x,t) = k^2 \Delta_x u(x,t) + F(x,t)$$
$$u(x,0) = 0 \qquad \frac{\partial}{\partial t} u(x,0) = 0$$

is solved by Duhamel's integral

$$u(x,t) = \int_0^t \left(\int_{\mathbb{R}^n} H_s(y) F(x-y, t-s)\, dy \right) ds.$$

where

$$H_s = \mathcal{F}^{-1}\left(\frac{\sin ks|\xi|}{k|\xi|} \right).$$

Show this is valid for negative t as well. Use this to solve the inhomogeneous wave equation with inhomogeneous initial data.

16. Interpret the solution in problem 15 in terms of finite propagation speed and Huyghens' principle ($n = 3$) for the influence of the inhomogeneous term $F(x,t)$.

17. Show that if the initial temperature is a radial function then the temperature at all later times is radial.

18. Maxwell's equations in a vacuum can be written

$$\frac{1}{c}\frac{\partial}{\partial t} E(x,t) = \text{curl } H(x,t) \qquad \text{div } E(x,t) = 0$$
$$\frac{1}{c}\frac{\partial}{\partial t} H(x,t) = -\text{curl } E(x,t) \qquad \text{div } H(x,t) = 0$$

where the electric and magnetic fields E and H are vector-valued functions on $\mathbb{R}^3$. Show that each component of these fields satisfies the wave equation with speed of propagation c.

19. Let $u(x,t)$ be a solution of the free Schrödinger equation with initial wave function φ satisfying $\int_{\mathbb{R}^3} |\varphi(x)|\, dx < \infty$. Show that $|u(x,t)| \leq ct^{-3/2}$ for some constant c. What does this tell you about the probabilty of finding a free particle in a bounded region of space as time goes by?

Chapter 6

The Structure of Distributions

6.1 The support of a distribution

The idea of *support* of an object (function, distribution, etc.) is the set on which it does something nontrivial. Roughly speaking, it is the complement of the set on which the object is zero, but this is not exactly correct. Consider, for example, the function $f(x) = x$ on the line. This function is zero at $x = 0$, and so at first we might be tempted to say that the support is $\mathbb{R}^1$ minus the origin. But this function is not completely dead at $x = 0$. It vanishes there, but not at nearby points. For a function to be truly boring at a point, it should vanish in a whole neighborhood of the point. Therefore, we define the *support* of a function f (written supp (f)) to be the *complement of the set of points x such that f vanishes in a neighborhood of x*. In our example, the support is the whole line. By the nature of the definition, the support is always a closed set, since it is the complement of an open set. For a continuous function, the support of f is just the closure of the set where f is different from zero. It follows that supp (f) is a *compact* set (recall that the compact sets in $\mathbb{R}^n$ are characterized by two conditions: they must be closed and bounded) if and only if f vanishes outside a bounded set. Such functions are said to have *compact support*. The test functions in $\mathcal{D}(\mathbb{R}^n)$ all have compact support. Thus we can describe the class of test functions succinctly as the C^∞ functions of compact support (the same is true if we consider test functions on an open set $\Omega \subset \mathbb{R}^n$).

Now we would like to make the same definition for distributions. We begin by defining what it means for a distribution to be identically zero in a neighborhood of a point. Since a neighborhood of a point just means an open set containing the point, it is the same thing to define $T \equiv 0$ on Ω for an open set Ω.

Definition 6.1.1 $T \equiv 0$ *on* Ω *means* $\langle T, \varphi \rangle = 0$ *for all test functions* φ *with* supp $\varphi \subseteq \Omega$. *Then* supp T *is the complement of the set of points* x *such that* T *vanishes in a neighborhood of* x.

Intuitively, if we cannot get a rise out of T by doing anything on Ω, then T is dead on Ω. However, it is important to understand what the definition does *not* say: since the support of φ is a closed set and Ω is an open set, the statement supp $\varphi \subseteq \Omega$ is stronger than saying φ vanishes at every point not in Ω; it says that φ vanishes in a neighborhood of every point not in Ω. For example, if $\Omega = (0, 1)$ in $\mathbb{R}^1$ then φ must vanish outside $(\epsilon, 1 - \epsilon)$ for some $\epsilon > 0$ in order that supp $\varphi \subseteq \Omega$.

To understand the definition, we need to look at some examples. Consider first the δ-function. Intuitively, the support of this distribution should be the origin. Indeed, if Ω is any open set not containing the origin, then $\delta \equiv 0$ on Ω, because $\langle \delta, \varphi \rangle = \varphi(0) = 0$ if supp $\varphi \subseteq \Omega$. On the other hand, we certainly do not have $\delta \equiv 0$ on a neighborhood of the origin since every such neighborhood supports test functions with $\varphi(0) = 1$. Thus supp $\delta = \{0\}$ as expected.

What is the support of $\delta'(n = 1)$? Suppose Ω is an open set not containing the origin, and supp $\varphi \subseteq \Omega$. Then φ vanishes *in a neighborhood of the origin* (not just at 0), and so $\langle \delta', \varphi \rangle = -\varphi'(0) = 0$. (In other words, $\varphi(0) = 0$ does not imply $\varphi'(0) = 0$, but $\varphi(x) = 0$ for all x in a neighborhood of 0 does imply $\varphi'(0) = 0$.) It is also easy to see that δ' does not vanish in any neighborhood of 0 (just construct φ supported in the neighborhood with $\varphi'(0) = 1$). Thus supp $\delta' = \{0\}$. Intuitively, both δ and δ' have the same support, but δ' "hangs out," at least infinitesimally, beyond its support. In other words, $\langle \delta', \varphi \rangle$ depends not just on the values of φ on the support $\{0\}$, but also on the values of the derivatives of φ on the support. This is true in general: the support of T is the smallest closed set E such that $\langle T, \varphi \rangle$ depends only on the values of φ and all its derivatives on E.

We can already use this concept to explain the finite propagation speed and Huyghens' principle for the wave equation discussed in section 5.3. The finite propagation speed says that the distributions $\mathcal{F}^{-1}(\cos kt|\xi|)$ and $\mathcal{F}^{-1}(\sin kt|\xi|/k|\xi|)$ are supported in the ball $|x| \leq kt$ (we say *supported in* when the support is known to be a subset of the given set), and Huyghens' principle says that the support is actually the sphere $|x| = kt$. Since supports add under convolution,

$$\text{supp } T * f \subseteq \text{supp } T + \text{supp } f$$

where the sum set $A + B$ is defined to be the set of all sums $a + b$ with $a \in A$ and $b \in B$.

Exercise: Verify this intuitively from the formula

$$T * f(x) = \int T(x - y)f(y)\, dy.$$

The properties of the solution

$$u(x, t) = \mathcal{F}^{-1}(\cos kt|\xi|) * f + \mathcal{F}^{-1}\left(\frac{\sin kt|\xi|}{k|\xi|}\right) * g$$

follow from these support statements.

As a more mundane example of the support of a distribution, consider the distribution $\langle T, \varphi \rangle = \int f(x)\varphi(x)\, dx$ associated to a continuous function f. If the definition is to be consistent, supp (T) should equal the support of f as a function. This is indeed the case, and the containment supp $(T) \subseteq$ supp (f) is easy to see from the contrapositive: if f vanishes in an open set Ω, then T vanishes there as well. Thus the complement of supp (f) is contained in the complement of supp (T). Conversely, if T vanishes on an open set Ω, then by considering test functions approximating a δ-function $\delta(x - y)$ for $y \in \Omega$ we can conclude that f vanishes at y.

A more interesting question arises if we drop the assumption that f be continuous. We still expect supp (T) to be equal to supp (f), but what do we mean by supp (f)? To understand why we have to be careful here, we need to look at a peculiar example. Let f be a function on the line that is zero for every x except $x = 0$, and let $f(0) = 1$. By our previous definition for continuous functions, the support of f should be the origin. But the distribution associated with f is the zero distribution, which has support equal to the empty set.

Of course this seems like a stupid example, since the same distribution could be given by the zero function, which would yield the correct support. Rather than invent some rule to bar such examples (actually, you won't succeed if you try!) it is better to reconsider what we mean by "f vanishes on Ω" if f is not necessarily continuous. What we want is that the integral of f should be zero on Ω, but not because of cancellation of positive and negative parts. That means we want

$$\int_\Omega |f(x)|\, dx = 0$$

as the condition for f to vanish on Ω. If this is so then

$$\int_\Omega f(x)\varphi(x)\, dx = 0$$

for every test function φ with support in Ω (in fact this is true for any function φ). This should seem plausible, and it follows from the inequalities

$$\left| \int_\Omega f(x)\varphi(x)\,dx \right| \leq \int_\Omega |f(x)|\,|\varphi(x)|\,dx$$

$$\leq M \int |f(x)|\,dx = 0$$

where M is the maximum value of $|\varphi(x)|$ (if φ is unbounded this requires a more complicated proof, but in the case at hand φ is a test function and so is bounded). So we have $f \equiv 0$ on Ω implies $T \equiv 0$ on Ω, and once again supp (T) = supp (f) if we define supp (f) as before (x is not in supp (f) if and only if f vanishes in a neighborhood of x.)

The distributions on $\mathbb{R}^n$ with compact support form an important class of distributions. Since the supp (T) is always closed, to say it is compact is the same thing as saying it is bounded. (For distributions on an open set Ω, compact support means that the support also stays away from the boundary of Ω.) The distributions of compact support are denoted $\mathcal{E}'(\mathbb{R}^n)$ (or $\mathcal{E}'$, for short), and this notation should make you suspect that they can also be considered as continuous linear functions on some space of test functions named $\mathcal{E}$. In fact it turns out that $\mathcal{E}$ is just the space of all C^∞ functions, with no restriction on the growth at infinity. What we are claiming is the following:

1. if T is a distribution of compact support, then $\langle T, \varphi \rangle$ makes sense for every C^∞ function φ; and

2. if T is a distribution, and if $\langle T, \varphi \rangle$ makes sense for every C^∞ function φ, then T must have compact support.

It is easy to explain why (1) is true. Suppose the support of T is contained in the ball $|x| \leq R$ (this will be true for some R since the support is bounded). Choose a function $\psi \in \mathcal{D}$ that is identically one on some larger ball. If φ is C^∞ then $\varphi\psi$ is in $\mathcal{D}$ and $\varphi\psi = \varphi$ on a neighborhood of the support of T. Thus it makes sense to define

$$\langle T, \varphi \rangle = \langle T, \psi\varphi \rangle$$

since T doesn't care what happens outside a neighborhood of its support. It is easy to see that the definition does not depend on the choice of ψ (a different choice ψ' would yield $\langle T, \psi'\varphi \rangle = \langle T, \psi\varphi \rangle + \langle T, (\psi' - \psi)\varphi \rangle$ and the last term is zero because $(\psi' - \psi)\varphi$ vanishes in a neighborhood of the support of T). We thus have associated a linear functional on $\mathcal{E}$ to each distribution of compact support, and it can be shown to be continuous in the required sense. The converse (2) is not as intuitive and requires the correct notion of continuity (see section 6.5). The idea is that if T did not have compact support we could construct a C^∞ function φ for which $\langle T, \varphi \rangle$ would have to be infinite, because no matter how quickly T tries to go to zero as $x \to \infty$, we can construct φ to grow even more rapidly as $x \to \infty$.

We can now consider all three spaces of test functions and distributions as a kind of hierarchy. For the test functions we have the containments

$$\mathcal{D} \subseteq \mathcal{S} \subseteq \mathcal{E}$$

(not alphabetical order, alas, even in French). For the distributions the containments are reversed:

$$\mathcal{E}' \subseteq \mathcal{S}' \subset \mathcal{D}'.$$

In particular, since distributions of compact support are tempered, the Fourier transform theory applies to them.

We will have a lot more to say about this in section 7.2.

6.2   Structure theorems

We have seen many distributions so far, but we have not said much about what the most general distribution looks like. We know that there are distributions that come from functions, and it is standard abuse of notation to say that they *are* the functions they come from. Since the space of distributions is closed under differentiation, there are all the derivatives of functions. It is almost true to say "that's all." If we view distribution theory as the completion of the differential calculus, then this would say that it is a minimal completion... we have not added anything we were not absolutely compelled to add.

But why did I say "almost" true? A more precise statement is that it is "locally" true. If T is a distribution of compact suppport, or even a tempered distribution, then it is true that T can be written as a finite sum of derivatives of functions. If T is only a distribution, then T can be written as a possibly infinite sum of derivatives of functions that is locally finite, meaning that for a fixed test function φ, only a finite number of terms are nonzero. For example, there is

$$T = \sum_{n=1}^{\infty} \delta^{(n)}(x - n),$$

that means

$$\langle T, \varphi \rangle = \sum_{n=1}^{\infty} \varphi^{(n)}(n).$$

For each fixed φ, all but a finite number of terms will be zero. Of course we have not yet written T as a sum of derivatives of functions, but recall that $\delta = H'$ where H is the Heaviside function

$$H(x) = \begin{cases} 1 & x > 0 \\ 0 & x \leq 0 \end{cases}$$

so

$$T = \sum_{n=1}^{\infty} H^{(n+1)}(x - n)$$

does the job. In fact, we can go a step further and make the functions continuous, if we note that $H = X'$ where

$$X(x) = \begin{cases} x & \text{if } x > 0 \\ 0 & \text{if } x \leq 0. \end{cases}$$

Thus we can also write

$$T = \sum_{n=1}^{\infty} X^{(n+2)}(x - n).$$

This is typical of what is true in general.

The results we are talking about can be thought of as "structure theorems" since they describe the structure of the distribution classes $\mathcal{E}', \mathcal{S}'$, and $\mathcal{D}'$. On the other hand, they are not as explicit as they might seem, since it is rather difficult to understand exactly what the 27th derivative of a complicated function is.

Before stating the structure theorems in $\mathbb{R}^n$, we need to introduce some notation. This *multi-index notation* will also be very useful when we discuss general partial differential operators. We use lowercase greek letters $\alpha, \beta, \ldots$ to denote multi-indices, which are just n-tuples $\alpha = (\alpha_1, \ldots, \alpha_n)$ of nonnegative integers. Then $(\partial/\partial x)^{\alpha}$ stands for $(\partial/\partial x_1)^{\alpha_1} \cdots (\partial/\partial x_n)^{\alpha_n}$. Similarly $x^{\alpha} = x_1^{\alpha_1} \cdots x_n^{\alpha_n}$. We write $|\alpha| = \alpha_1 + \cdots + \alpha_n$ and call this the *order* of the multi-index. This is consistent with the usual definition of the order of a partial derivative. We also write $\alpha! = \alpha_1! \cdots \alpha_n!$.

Structure Theorem for $\mathcal{E}'$: Let T be a distribution of compact support on $\mathbb{R}^n$. Then there exist a finite number of continuous functions f_{α} such that $T = \sum (\partial/\partial x)^{\alpha} f_{\alpha}$ (finite sum), and each term $(\partial/\partial x)^{\alpha} f_{\alpha}$ has compact support.

Structure Theorem for $\mathcal{S}'$: Let T be a tempered distribution on $\mathbb{R}^n$. Then there exist a finite number of continuous functions $f_\alpha(x)$ satisfying a polynomial order growth estimate

$$|f_\alpha(x)| \le c(1 + |x|)^N$$

such that

$$T = \sum \left(\frac{\partial}{\partial x}\right)^\alpha f_\alpha \quad \text{(finite sum)}.$$

Structure Theorem for $\mathcal{D}'$: Let T be a distribution on $\mathbb{R}^n$. Then there exist continuous functions f_α such that

$$T = \sum \left(\frac{\partial}{\partial x}\right)^\alpha f_\alpha \quad \text{(infinite sum)}$$

where for every bounded open set Ω, all but a finite number of the distributions $(\partial/\partial x)^\alpha f_\alpha$ vanish identically on Ω.

There is no uniqueness corresponding to these representations of T. In the first two structure theorems you can even consolidate the finite sum into a single term, $T = (\partial/\partial x)^\alpha f$ for some α and f. For example, if $n = 1$ and, say, $T = f_0 + f_1'$, then also $T = f'$ where $f(x) = f_1(x) + \int_0^x f_0(y) \, dy$.

Exercise: Verify this using the fundamental theorem of the calculus. The trouble with this consolidation is that, when $n > 1$, it may increase the order of the derivatives. For example, to consolidate

$$\frac{\partial}{\partial x_1} f_1 + \frac{\partial}{\partial x_2} f_2$$

requires $(\partial^2/\partial x_1 \partial x_2)f$.

The highest order of derivative (the maximum of $|\alpha| = \alpha_1 + \alpha_2 + \cdots + \alpha_n$) in a representation $T = \sum (\partial/\partial x)^\alpha f_\alpha$ is called the *order* of the distribution. The trouble with this definition is that we really want to take the minimum value of this order over all such representations (remember the lack of uniqueness). Thus if we know $T = \sum (\partial/\partial x)^\alpha f_\alpha$ with $|\alpha| \le m$ then we can only assert that the order of T is $\le m$, since there might be a better representation. (Also, there is no general agreement on whether or not to insist that the functions f_α in the representation be continuous; frequently one obtains such representations where the functions f_α are only locally bounded, or locally integrable, and for many purposes that is sufficient.) However, despite all these caveats, the imprecise notion of the order of a distribution is a widely used concept. One conclusion that is beyond question is that the first two structure theorems assert that every distribution in

$\mathcal{E}'$ or $\mathcal{S}'$ has finite order. We have seen above an example of a distribution of infinite order.

One subtle point in the first structure theorem is that the functions f_α themselves need not have compact support. For example, the δ-function is not the derivative (of any order) of a function of compact support, even though δ has support at the origin. The reason for this is that $\langle T', 1 \rangle = 0$ if T has compact support, where 1 denotes the constant function. Indeed $\langle T', 1 \rangle = \langle T', \psi \rangle$ by definition, where $\psi \in \mathcal{D}$ is one on a neighborhood of the support of T'. But $\langle T', \psi \rangle = \langle T, \psi' \rangle$ and ψ' vanishes on a neighborhood of the support of T, so $\langle T', \psi \rangle = 0$. Notice that this argument does not work if T' has compact support but T does not.

The proofs of the structure theorems are not elementary. There are some proofs that are short and tricky and entirely nonconstructive. There are constructive proofs, but they are harder. When $n = 1$, the idea behind the proof is that you integrate the distribution many times until you obtain a continuous function. Then that function differentiated as many times brings you back to the original distribution. The problem is that it is tricky to define the integral of a distribution (see problem 9). Once you have the right definition, it requires the correct notion of continuity of a distribution to prove that for T in $\mathcal{E}'$ or $\mathcal{S}'$ you eventually arrive at a continuous function after a finite number of integrals (of course this may not be true for general distributions).

One aspect of the proofs that is very simple is that the structure theorem for $\mathcal{D}'$ follows easily from the structure theorem for $\mathcal{E}'$. The idea uses the important idea of a *partition of unity*. This is by definition a sequence of test functions $\varphi_j \in \mathcal{D}$ such that $\sum_j \varphi_j \equiv 1$ and the sum is locally finite. It is easy to construct partitions of unity (see problem 10), and for an arbitrary distribution T we then have $T = \sum_j \varphi_j T$ (locally finite sum). Note that $\varphi_j T$ has compact support (the support of φ_j). If we apply the first structure theorem to each $\varphi_j T$ and sum, we obtain the third structure theorem.

6.3   Distributions with point support

We have seen that the δ-function and all its derivatives have support equal to the single point $\{0\}$. In view of the structure theorems of the last section, it is tempting to guess that there are no others; a distribution with support $\{0\}$ must be of the form

$$T = \sum a_\alpha \left(\frac{\partial}{\partial x} \right)^\alpha \delta \quad \text{(finite sum)}.$$

On the other hand, we might also conjecture that certain infinite sums might be allowed if the coefficients a_α tend to zero rapidly enough. It turns out that the first guess is right, and the second is wrong. In fact, the following is true: given any numbers c_α, there exists a test function φ in $\mathcal{D}$ such that $(\partial/\partial x)^\alpha \varphi(0) = c_\alpha$. In other words, there is no restriction on the Taylor expansion of a C^∞ function (recall that the Taylor expansion about 0 is

$$\sum \frac{1}{\alpha!} \left(\frac{\partial}{\partial x}\right)^\alpha \varphi(0) x^\alpha).$$

This means that the Taylor expansion does not have to converge (even if it does converge, it does not have to converge to the function, unless φ is analytic).

Suppose we tried to define $T = \sum a_\alpha (\partial/\partial x)^\alpha \delta$ where infinitely many a_α are nonzero. Then $\langle T, \varphi \rangle = \sum (-1)^{|\alpha|} a_\alpha c_\alpha$ and by choosing c_α appropriately we can make this infinite. So such infinite sums of derivatives of δ are not distributions. Incidentally, it is not hard to write down a formula for φ given c_α. It is just

$$\varphi(x) = \sum \frac{c_\alpha}{\alpha!} x^\alpha \psi(\lambda_\alpha x)$$

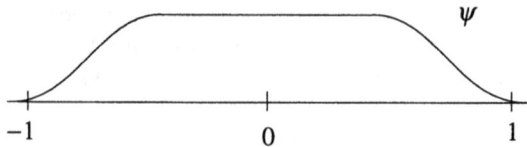

Figure 6.1

where $\psi \in \mathcal{D}$ is equal to one in a neighborhood of the origin, and $\lambda_\alpha \to \infty$ rapidly as $\alpha \to \infty$. If supp ψ is $|x| \leq 1$, then supp $\psi(\lambda_\alpha x)$ is $|x| \leq \lambda_\alpha^{-1}$ which is shrinking rapidly. For any fixed $x \neq 0$, only a finite number of terms in the sum defining φ will be nonzero, and for $x = 0$ only the first term is nonzero. If we are allowed to differentiate the series term by term then we get $(\partial/\partial x)^\alpha \varphi(0) = c_\alpha$ because all the derivatives of $\psi(\lambda_\alpha x)$ vanish at the origin. It is in justifying the term-by-term differentiation that all the hard work lies, and in fact the rate at which λ_α must grow will be determined by the size of the coefficients c_α.

Just because there are no infinite sums $\sum a_\alpha (\partial/\partial x)^\alpha \delta$ does not in itself prove that every distribution supported at the origin is a finite sum of this sort. It is conceivable that there could be other, wilder distributions, with one point support. In fact, there are not. Although I cannot give the entire proof, I can indicate some of the ideas involved. If T has support $\{0\}$, then $\langle T, \varphi \rangle = 0$ for every test function φ vanishing in a neighborhood of the origin. However, it then follows that $\langle T, \varphi \rangle = 0$ for every test function φ that vanishes to order N at zero, $(\partial/\partial x)^\alpha \varphi(0) = 0$ for all $|\alpha| \leq N$ (here N depends on the distribution T). This is a consequence of the structure theorem, $T = \sum (\partial/\partial x)^\alpha f_\alpha$ (finite sum) and an approximation argument: if φ vanishes to order N we approximate φ by $\varphi(x)(1 - \psi(\lambda x))$ as $\lambda \to \infty$. Then $\langle T, \varphi(x)(1 - \psi(\lambda x)) \rangle = 0$ because $\varphi(x)(1 - \psi(\lambda x))$ vanishes in a neighborhood of zero and

$$
\begin{aligned}
&\langle T, \varphi(x)(1 - \psi(\lambda x)) \rangle \\
&= \left\langle \sum \left(\frac{\partial}{\partial x}\right)^\alpha f_\alpha, \varphi(x)(1 - \psi(\lambda x)) \right\rangle \\
&= \sum (-1)^{|\alpha|} \int f_\alpha(x) \left(\frac{\partial}{\partial x}\right)^\alpha [\varphi(x)(1 - \psi(\lambda x))]\, dx
\end{aligned}
$$

which converges as $\lambda \to \infty$ to

$$
\sum (-1)^{|\alpha|} \int f_\alpha(x) \left(\frac{\partial}{\partial x}\right)^\alpha \varphi(x)\, dx = \langle T, \varphi \rangle
$$

(the convergence as $\lambda \to \infty$ follows from the fact that φ vanishes to order N at zero and we never take more than N derivatives—the value of N is the upper bound for $|\alpha|$ in the representation $T = \sum (\partial/\partial x)^\alpha f_\alpha$). Thus $\langle T, \varphi \rangle$ is the limit of terms that are always zero, so $\langle T, \varphi \rangle = 0$. This is the tricky part of the argument, because for general test functions, say if $\varphi(0) \neq 0$, the approximation of $\varphi(x)$ by $\varphi(x)(1 - \psi(\lambda x))$ is no good—it gives the wrong value at $x = 0$.

Once we know that $\langle T, \varphi \rangle = 0$ for all φ vanishing to order N at the origin, it is just a matter of linear algebra to complete the proof. Let us write, temporarily, $\mathcal{D}_N$ for this space of test functions. It is then easy to construct a finite set of test functions φ_α (one for each α with $|\alpha| \leq N$) so that every test function φ can be written uniquely $\varphi - \tilde{\varphi} + \sum c_\alpha \varphi_\alpha$ with $\tilde{\varphi} \subset \mathcal{D}_N$. Indeed we just take $\varphi_\alpha = (1/\alpha!) x^\alpha \psi(x)$, and then $c_\alpha = (\partial/\partial x)^\alpha \varphi(0)$ does the trick. In terms of linear algebra, $\mathcal{D}_N$ has finite codimension in $\mathcal{D}$. Now $\langle T, \varphi \rangle = \sum c_\alpha \langle T, \varphi_\alpha \rangle$ for any $\varphi \in \mathcal{D}$ because $\langle T, \tilde{\varphi} \rangle = 0$. But the distribution $\sum b_\alpha (\partial/\partial x)^\alpha \delta$ satisfies

$$
\begin{aligned}
\left\langle \sum b_\alpha \left(\frac{\partial}{\partial x}\right)^\alpha \delta, \varphi \right\rangle &= \sum \sum c_\beta b_\alpha \left\langle \left(\frac{\partial}{\partial x}\right)^\alpha \delta, \varphi_\beta \right\rangle \\
&= \sum (-1)^{|\alpha|} c_\alpha b_\alpha
\end{aligned}
$$

since

$$\left\langle \left(\frac{\partial}{\partial x}\right)^\alpha \delta, \varphi_\beta \right\rangle = 0$$

if $\alpha \neq \beta$ and 1 if $\alpha = \beta$. Thus we need only choose $b_\alpha = (-1)^{|\alpha|} \langle T, \varphi \rangle$ in order to have $\langle T, \varphi \rangle$ equal to

$$\left\langle \sum b_\alpha \left(\frac{\partial}{\partial x}\right)^\alpha \delta, \varphi \right\rangle$$

for every test function φ, or $T = \sum b_\alpha (\partial/\partial x)^\alpha \delta$.

We have already seen an interesting application of this theorem: if a harmonic function has polynomial growth, then it must be a polynomial. If $|u(x)| \leq c(1 + |x|)^N$ then u may be regarded as a tempered distribution. Then $\Delta u = 0$ means $-|\xi|^2 \hat{u}(\xi) = 0$ which easily implies supp $\hat{u} = \{0\}$ (see problem 16 for a more general result). So $\hat{u}(\xi) = \sum c_\alpha (\partial/\partial \xi)^\alpha \delta$ which implies u is a polynomial. Of course not every polynomial is harmonic, but for $n \geq 2$ there is an interesting theory of harmonic polynomials called *spherical harmonics*.

6.4 Positive distributions

Mathematicians made an unfortunate choice when they decided that *positive* should mean strictly greater than zero, leaving the unwieldly term *nonnegative* to mean ≥ 0. That makes liars of us, because we like to say *positive* when we really mean *nonnegative*. Thus we will define a *positive distribution* to be one that takes nonnegative values on nonnegative test functions, $\langle T, \varphi \rangle \geq 0$ if $\varphi \geq 0$. (Throughout this section we deal only with real-valued functions and distributions.) Clearly, if $\langle T, \varphi \rangle = \int f(x)\varphi(x)\, dx$ and $f \geq 0$ then T is a positive distribution. Moreover, if f is continuous, then T is a positive distribution if and only if $f \geq 0$. If f is not continuous then we encounter the same sort of difficulty that we did in the discussion of the support of f. The resolution of the difficulty is similar: if we write $f = f^+ - f^-$ as the unique decomposition into positive and negative parts (so $f^+(x) = \max(f(x), 0)$ and $f^-(x) = \max(-f(x), 0)$) then T is a positive distribution if and only if $\int f^-(x)\, dx = 0$.

Thus the definition of positive distribution works the way it should for functions. Are there any other positive distributions? You bet! The δ-function is positive, since $\langle \delta, \varphi \rangle = \varphi(0) \geq 0$ if $\varphi \geq 0$. But δ' is not positive, (nor is $-\delta'$) since we cannot control the sign of the derivative by controlling the sign of the function.

A natural question arises whenever we define positivity for any class of objects: is everything representable as a difference of two positive objects?

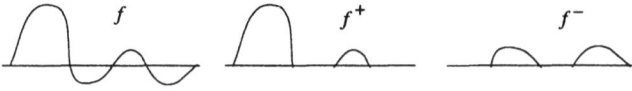

Figure 6.2

If so, we can ask if there is some canonical choice of positive and negative parts. But for distributions we never get to the second question, because the answer to the first is a resounding NO! Very few distributions can be written in the form $T_1 - T_2$ for T_1 and T_2 positive distributions. In particular, δ' cannot.

To understand why positive distributions are so special we need to turn the positivity condition into a slightly different, and considerably stronger, inequality. It is easier to present the argument for distributions of compact support. So we suppose supp (T) is contained in $|x| \leq R$, and we choose a nonnegative test function ψ that is identically one on a neighborhood of supp (T). Then $\langle T, \psi \rangle$ is a positive number; call it M. We then claim that $|\langle T, \varphi \rangle| \leq cM$ where $c = \sup_x |\varphi(x)|$, for any test function φ.

To prove this claim we simply note that the test functions $c\psi \pm \varphi$ are positive on a neighborhood of supp (T), and so $\langle T, c\psi \pm \varphi \rangle \geq 0$ (here we have used the observation that if T is a positive distribution of compact support we only need nonnegativity of the test function φ on a neighborhood of the support of T to conclude $\langle T, \varphi \rangle \geq 0$). But then both $\langle T, \varphi \rangle \leq \langle T, c\psi \rangle$ and

$$-\langle T, \varphi \rangle \leq \langle T, c\psi \rangle$$

which is the same as

$$|\langle T, \varphi \rangle| \leq \langle T, c\psi \rangle = cM$$

as claimed.

What does this inequality tell us? It says that the size of $\langle T, \varphi \rangle$ is controlled by the size of φ (the maximum value of $|\varphi(x)|$). That effectively rules out anything that involves derivatives of φ, because φ can wiggle a lot (have large derivatives) while remaining small in absolute value.

Now suppose $T = T_1 - T_2$ where T_1 and T_2 are positive distributions of compact support. Then if M_1 and M_2 are the associated constants we have

$$|\langle T, \varphi \rangle| = |\langle T_1, \varphi \rangle - \langle T_2, \varphi \rangle| \leq |\langle T_1, \varphi \rangle| + |\langle T_2, \varphi \rangle| \leq c(M_1 + M_2)$$

so T satisfies the same kind of inequality. Since we know δ' does not satisfy such an inequality, regardless of the constant, we have shown that it is not the difference of positive distributions of compact support. But it is easy to localize the argument to show that $\delta' \neq T_1 - T_2$ even if T_1 and T_2 do not have compact support: just choose ψ as before and observe that $\delta' = T_1 - T_2$ would imply $\delta' = \psi\delta' = \psi T_1 - \psi T_2$ and ψT_1 and ψT_2 would be positive distributions of compact support after all.

For readers who are familiar with measure theory, I can explain the story more completely. The positive distributions all come from positive measures. If μ is a positive measure that is locally finite ($\mu(A) < \infty$ for every bounded set A), then $\langle T, \varphi \rangle = \int \varphi \, d\mu$ defines a positive distribution. The converse is also true: every positive distribution comes from a locally finite positive measure. This follows from a powerful result known as the Riesz representation theorem and uses the inequality we derived above. If you are not familiar with measure theory, you should think of a positive measure as the most general kind of probability distribution, but without the assumption that the total probabity adds up to one.

For an example of an "exotic" probability measure consider the usual construction of the Cantor set: take the unit interval, delete the middle third, then in each of the remaining intervals delete the middle third, and iterate the deletion process infinitely often.

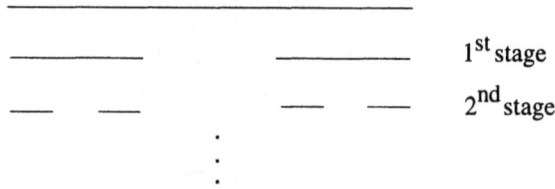

Figure 6.3

Now define a probability measure by saying each of the two intervals in stage 1 has probability $\frac{1}{2}$, each of the four intervals in stage 2 has probability $\frac{1}{4}$, and in general each of the 2^n intervals in the nth stage has probability 2^{-n}. The resulting measure is called the *Cantor measure*. It assigns total measure one to the Cantor set and zero to the complement. The usual length measure of the Cantor set is zero.

6.5   Continuity of distribution

So far I have been deliberately vague about the meaning of continuity in the definition of distribution, and I have described for you most of the theory of distribution without involving continuity. For a deeper understanding of the theory you should know what continuity means, and it may also help to clarify some of the concepts we have already discussed. There are at least three equivalent ways of stating the meaning of continuity, namely

1. a small change in the input yields a small change in the output,

2. interchanging limits, and

3. estimates.

What does this mean in the context of distribution? We want to think of a distribution T as a function on test functions, $\langle T, \varphi \rangle$. So (1) would say that if φ_1 is close to φ_2 then $\langle T, \varphi_1 \rangle$ is close to $\langle T, \varphi_2 \rangle$, (2) would say that if $\lim_{j \to \infty} \varphi_j = \varphi$ then $\lim_{j \to \infty} \langle T, \varphi_j \rangle = \langle T, \varphi \rangle$, and (3) would say that $|\langle T, \varphi \rangle|$ is less than a constant times some quantity that measures the size of φ. Statements like (1) and (2) should be familiar from the definition of continuous function, although we have not yet explained what "φ_1 is close to φ_2" or "$\lim_{j \to \infty} \varphi_j = \varphi$" should mean. Statement (3) has no analog for continuity of ordinary functions, and depends on the linearity of $\langle T, \varphi \rangle$ for its success. Readers who have seen some functional analysis will recognize this kind of estimate, which goes under the name "boundedness."

To make (1) precise we now list the conditions we would like to have in order to say that φ_1 is close to φ_2. We certainly want the values $\varphi_1(x)$ close to $\varphi_2(x)$, and this should hold uniformly in x. To express this succinctly it is convenient to introduce the "sup-norm" notation:

$$\|\varphi\|_\infty = \sup_x |\varphi(x)|.$$

The meaning of "norm" is that

$$\|\varphi_1 + \varphi_2\|_\infty \le \|\varphi_1\|_\infty + \|\varphi_2\|_\infty \text{ and } \|a\varphi\|_\infty = |a|\,\|\varphi\|_\infty$$

and $0 \le \|\varphi\|_\infty$ with equality if and only if $\varphi \equiv 0$.

Exercise: Verify these conditions.

So we want $\|\varphi_1 + \varphi_2\|_\infty \le \epsilon$ if φ_1 is close to φ_2. But we have to demand more, because in our work with test functions we often differentiate them. Thus we will demand also

$$\left\| \left(\frac{\partial}{\partial x} \right)^\alpha \varphi_1 - \left(\frac{\partial}{\partial x} \right)^\alpha \varphi_2 \right\|_\infty \le \epsilon,$$

which is not a consequence of $||\varphi_1 + \varphi_2||_\infty \leq \epsilon$ even if we change ϵ (a small wiggly function φ may be close to zero in the sense that $||\varphi - 0||_\infty = ||\varphi||_\infty$ is small, but $||\varphi'||_\infty$ may be very large). To be more precise, we can only demand

$$\left\|\left(\frac{\partial}{\partial x}\right)^\alpha \varphi_1 - \left(\frac{\partial}{\partial x}\right)^\alpha \varphi_2\right\|_\infty \leq \epsilon$$

for a finite number of derivatives, because higher order derivatives tend to grow rapidly as the order of the derivative goes to infinity.

There is one more condition we will require in order to say φ_1 is close to φ_2, and this condition is not something you might think of at first: we want the supports of φ_1 and φ_2 to be fairly close. Actually it will be enough to demand that the supports be contained in a fixed bounded set, say $|x| \leq R$ for some given R. Without such a condition, we would have to say that a test function φ is close to 0 even if it has a huge support (of course it would have to be very small far out), and this would be very awkward for the kind of locally finite infinite sums we considered in section 6.2.

So, altogether, we will say that φ_1 is close to φ_2 if

$$\left\|\left(\frac{\partial}{\partial x}\right)^\alpha \varphi_1 - \left(\frac{\partial}{\partial x}\right)^\alpha \varphi_2\right\|_\infty \leq \delta$$

for all $|\alpha| \leq m$ and supp φ_1, supp $\varphi_2 \subseteq \{|x| \leq R\}$. Of course "close" is a qualitative word, but the three parameters δ, m, R give it a quantitative meaning. The statement that T is *continuous* at φ_1 is the statement: *for every $\epsilon > 0$ and R sufficiently large there exist parameters δ, m (depending on φ_1, R, and ϵ) such that if φ_1 is close to φ_2 with parameters δ, m, R then*

$$|\langle T, \varphi_1 \rangle - \langle T, \varphi_2 \rangle| \leq \epsilon.$$

Continuity of T is defined to be continuity at every test function. Notice that this really says we can make $\langle T, \varphi_1 \rangle$ close to $\langle T, \varphi_2 \rangle$ if we make φ_1 close to φ_2, but we must make φ_1 close to φ_2 in the rather complicated way described above.

Now we can describe the second definition of continuity in similar terms. We need to explain what we mean by $\lim_{j\to\infty} \varphi_j = \varphi$ for test functions. It comes as no surprise that we want

$$\lim_{j\to\infty} \left\|\left(\frac{\partial}{\partial x}\right)^\alpha \varphi_j - \left(\frac{\partial}{\partial x}\right)^\alpha \varphi\right\|_\infty = 0$$

for every α ($\alpha = 0$ takes care of the case of no derivatives), and the condition on the supports is that there exists a bounded set, say $|x| \leq R$, that contains supp φ and all supp φ_j. Again this is not an obvious condition, but it is necessary if we are to have everything localized when we take limits. This is the definition of the limit process for $\mathcal{D}$. Readers who are familiar with the concepts will recognize that this makes $\mathcal{D}$ into a topological space, but not a metric space—there is no single notion of distance in any of our test function spaces $\mathcal{D}, \mathcal{S}$, or $\mathcal{E}$.

With this definition of limits of test functions, it is easy to see that if T is continuous then $\langle T, \varphi \rangle = \lim_{j \to \infty} \langle T, \varphi_j \rangle$ whenever $\varphi = \lim_{j \to \infty} \varphi_j$. The argument goes as follows: we use the continuity of T at φ. Given $\epsilon > 0$ we take R as above and then find δ, m so that $|\langle T, \varphi \rangle - \langle T, \psi \rangle| \leq \epsilon$ whenever φ and ψ are close with parameters δ, m, R. Then every φ_j has support in $|x| \leq R$, and for j large enough

$$\left\| \left(\frac{\partial}{\partial x} \right)^{\alpha} \varphi_j - \left(\frac{\partial}{\partial x} \right)^{\alpha} \varphi \right\|_{\infty} \leq \delta$$

for all $|\alpha| \leq m$. This is the meaning of

$$\lim_{j \to \infty} \left\| \left(\frac{\partial}{\partial x} \right)^{\alpha} \varphi_j - \left(\frac{\partial}{\partial x} \right)^{\alpha} \varphi \right\|_{\infty} = 0.$$

Thus, for j large enough, all φ_j are close to φ with parameters δ, m, R and so

$$|\langle T, \varphi_j \rangle - \langle T, \varphi \rangle| \leq \epsilon.$$

But this is what $\lim_{j \to \infty} \langle T, \varphi_j \rangle = \langle T, \varphi \rangle$ means. A similar argument from the contrapositive shows the converse: if T satisfies $\lim_{j \to \infty} \langle T, \varphi_j \rangle = \langle T, \varphi \rangle$ whenever $\lim_{j \to \infty} \varphi_j = \varphi$ (in the above sense) then T is continuous.

This form of continuity is extremely useful, since it allows us to take limiting processes inside the action of the distribution. For example, if $\varphi_s(t)$ is a test function of the two variables (s, t), then

$$\frac{d}{ds} \langle T, \varphi_s \rangle = \left\langle T, \frac{\partial}{\partial s} \varphi_s \right\rangle.$$

The reason for this is first that the linearity allows us to bring the difference quotient inside the distribution,

$$\frac{1}{h} \left(\langle T, \varphi_{s+h} \rangle - \langle T, \varphi_s \rangle \right) = \left\langle T, \frac{1}{h} (\varphi_{s+h} - \varphi_s) \right\rangle$$

and then the continuity lets us take the limit as $h \to 0$ because

$$\lim_{n \to 0} \frac{1}{h} (\varphi_{s+h} - \varphi_s) = \frac{\partial}{\partial s} \varphi_s$$

in the sense of $\mathcal{D}$ convergence. Similarly with integrals:

$$\int_{-\infty}^{\infty} \langle T, \varphi_s \rangle \, ds = \left\langle T, \int_{-\infty}^{\infty} \varphi_s \, ds \right\rangle,$$

because the Riemann sums approximating $\int_{-\infty}^{\infty} \varphi_s \, ds$ converge in the sense of $\mathcal{D}$, and the Riemann sums are linear. We will use this observation in section 7.2 when we discuss the Fourier transform of distributions of compact support.

The third form of continuity, the *boundedness* of T, involves estimates that generalize the inequality

$$|\langle T, \varphi \rangle| \leq M \|\varphi\|_\infty$$

that a positive distribution of compact support must satisfy. The generalization involves allowing derivatives on the right side and also restricting the support of φ. Here is the exact statement: *for every R there exists M and m such that*

$$|\langle T, \varphi \rangle| \leq M \sum_{|\alpha| \leq M} \left\| \left(\frac{\partial}{\partial x} \right)^\alpha \varphi \right\|_\infty$$

for all φ with support in $|x| \leq R$. It is easy to see that this boundedness implies continuity in the first two forms. For example, if φ_1 and φ_2 both have support in $|x| \leq R$ then

$$
\begin{aligned}
|\langle T, \varphi_1 \rangle - \langle T, \varphi_2 \rangle| &= |\langle T, \varphi_1 - \varphi_2 \rangle| \\
&\leq M \sum_{|\alpha| \leq M} \left\| \left(\frac{\partial}{\partial x} \right)^\alpha \varphi_1 - \left(\frac{\partial}{\partial x} \right)^\alpha \varphi_2 \right\|_\infty,
\end{aligned}
$$

so if

$$\left\| \left(\frac{\partial}{\partial x} \right)^\alpha \varphi_1 - \left(\frac{\partial}{\partial x} \right)^\alpha \varphi_2 \right\|_\infty \leq \delta$$

for all $|\alpha| \leq m$ then

$$|\langle T, \varphi_1 \rangle - \langle T, \varphi_2 \rangle| \leq c_m M \delta$$

where c_m is the number of multi-indices α with $|\alpha| \leq m$. Since c_m and M are known in advance, we just take $\delta = \epsilon / c_m M$.

But conversely, we can show that continuity in the first sense implies boundedness by using the linearity of T in one of the basic principles of functional analysis. The continuity of T at the zero function means that given any $\epsilon > 0$, say $\epsilon = 1$, and R, there exist δ and m such that if supp $\varphi \subseteq \{|x| \leq R\}$ and $||(\partial/\partial x)^\alpha \varphi - 0||_\infty \leq \delta$ for all $|\alpha| \leq m$ then $|\langle T, \varphi \rangle| \leq 1$. Now the trick is that for any φ with supp $\varphi \subseteq \{|x| \leq R\}$ we can multiply it by a small enough constant c so that we have $||(\partial/\partial x)^\alpha c\varphi||_\infty \leq \delta$ for all $|\alpha| \leq m$. Just take c to be the smallest of $\delta ||(\partial/\partial x)^\alpha \varphi||_\infty^{-1}$ for $|\alpha| \leq m$. Then $|\langle T, c\varphi \rangle| \leq 1$ or $|\langle T, \varphi \rangle| \leq c^{-1}$ and

$$c^{-1} = \max_{|\alpha| \leq m} \delta^{-1} \left|\left|\left(\frac{\partial}{\partial x}\right)^\alpha \varphi\right|\right|_\infty \leq \delta^{-1} \sum_{|\alpha| \leq m} \left|\left|\left(\frac{\partial}{\partial x}\right)^\alpha \varphi\right|\right|_\infty$$

so we have the required boundedness condition with $M = \delta^{-1}$.

The boundedness form of continuity explains why positive distributions are so special—they satisfy the boundedness estimate with $m = 0$, or no derivatives. More generally, the number of derivatives involved in the boundedness estimate is related to the number of derivatives in the structure theorem. Recall that for $T \in \mathcal{D}'$ we had $T = \sum (\partial/\partial x)^\alpha f_\alpha$ for f_α continuous, where only a finite number of terms are nonzero on any bounded set. Suppose that on $|x| \leq R$ we only have terms with $|\alpha| \leq m$. In other words, if φ is supported in $|x| \leq R$, then

$$\langle T, \varphi \rangle = \sum_{|\alpha| \leq m} \left\langle \left(\frac{\partial}{\partial x}\right)^\alpha f_\alpha, \varphi \right\rangle$$

$$= \sum_{|\alpha| \leq m} (-1)^{|\alpha|} \int f_\alpha(x) \left(\frac{\partial}{\partial x}\right)^\alpha \varphi(x) \, dx.$$

But we have the estimate

$$\left| \int f_\alpha(x) \left(\frac{\partial}{\partial x}\right)^\alpha \varphi(x) \, dx \right| \leq \left|\left|\left(\frac{\partial}{\partial x}\right)^\alpha \varphi\right|\right|_\infty \int_{|x| \leq R} |f_\alpha(x)| \, dx$$

hence

$$|\langle T, \varphi \rangle| \leq M \sum_{|\alpha| \leq m} \left|\left|\left(\frac{\partial}{\partial x}\right)^\alpha \varphi\right|\right|_\infty$$

where M is the maximum value of

$$\int_{|x| \leq R} |f_\alpha(x)| \, dx$$

(since f_α is continuous, these integrals are all finite).

In fact, what we have just argued is that the structure theorem implies boundedness, which is putting the cart before the horse. In the proof of the structure theorem, the boundedness is used in a crucial way. However, the above converse argument at least makes the structure theorem plausible.

There are similar descriptions of continuity for the two other classes of distributions, $\mathcal{S}'$ and $\mathcal{E}'$, and in fact they are somewhat simpler to state because they do not involve supports. I will give the definition in terms of boundedness, and leave the first two continuity forms as an exercise. For distributions of compact support, the boundedness condition is just that there exist one estimate

$$|\langle T, \varphi \rangle| \leq M \sum_{|\alpha| \leq m} \left\| \left(\frac{\partial}{\partial x} \right)^{\alpha} \varphi \right\|_{\infty}$$

for some m and M, for all test functions φ, regardless of support. For tempered distributions, the boundedness condition is again just a single estimate, but this time of the form

$$|\langle T, \varphi \rangle| \leq M \sum_{|\alpha| \leq m} \left\| (1 + |x|)^{N} \left(\frac{\partial}{\partial x} \right)^{\alpha} \varphi(x) \right\|_{\infty}$$

for some m, M, and N, and all test functions (it turns out to be the same whether you require $\varphi \in \mathcal{D}$ or just $\varphi \in \mathcal{S}$, for reasons that will be discussed in the next section). This kind of condition is perhaps not so surprising if you recall that the finiteness of

$$\left\| (1 + |x|)^{N} \left(\frac{\partial}{\partial x} \right)^{\alpha} \varphi(x) \right\|_{\infty}$$

for all N and α is the defining property of $\mathcal{S}$.

So why didn't I tell you about the continuity conditions, at least in the form of boundedness, right from the start? It is usually not difficult to establish the boundedness estimate for any particular distribution. I claim that essentially all you have to do is examine the proof that the distribution is defined for all test functions, and throw in some obvious estimates. (I'll illustrate this in a moment.) But it is sometimes burdensome, and what's more, it makes the whole exposition of the theory burdensome. For example, when we defined $(\partial / \partial x_j) T$ by

$$\left\langle \frac{\partial}{\partial x_j} T, \varphi \right\rangle = - \left\langle T, \frac{\partial}{\partial x_j} \varphi \right\rangle,$$

we would have been required to show that $(\partial / \partial x_j) T$ is also bounded. That argument looks like this: since T is bounded, for every R there exists m and M such that

$$|\langle T, \varphi \rangle| \leq M \sum_{|\alpha| \leq m} \left\| \left(\frac{\partial}{\partial x} \right)^{\alpha} \varphi \right\|_{\infty}$$

if supp $\varphi \subseteq \{|x| \leq R\}$. For the same R, then, we have

$$\left| \left\langle \frac{\partial}{\partial x_j} T, \varphi \right\rangle \right| = \left| \left\langle T, \frac{\partial}{\partial x_j} \varphi \right\rangle \right| \leq M \sum_{|\alpha| \leq m} \left\| \left(\frac{\partial}{\partial x} \right)^\alpha \frac{\partial}{\partial x_j} \varphi \right\|_\infty$$

$$\leq M \sum_{|\alpha| \leq m+1} \left\| \left(\frac{\partial}{\partial x} \right)^\alpha \varphi \right\|_\infty$$

if supp $\varphi \subseteq \{|x| \leq R\}$, which is of the required form with m increased to $m + 1$. The same goes for all the other operations on distributions. Thus, by postponing the discussion of boundedness until now, I have saved both of us a lot of trouble.

Here is an example of how the boundedness is implicit in the proof of existence. Remember the distribution

$$\langle T, \varphi \rangle = \int_{-\infty}^{-1} + \int_1^\infty \frac{\varphi(x)}{|x|} \, dx + \int_{-1}^1 \frac{\varphi(x) - \varphi(0)}{|x|} \, dx$$

that was concocted out of the function $1/|x|$ that is not locally integrable? To show that it was defined for every test function φ, you had to invoke the mean value theorem,

$$\varphi(x) - \varphi(0) = x \varphi'(y)$$

for some y between 0 and x. Since y depends on x we will write it explicitly as $y(x)$. Then

$$\langle T, \varphi \rangle = \int_{-\infty}^{-1} + \int_1^\infty \frac{|\varphi(x)|}{|x|} \, dx + \int_{-1}^1 \varphi'(y(x)) \operatorname{sgn} x \, dx$$

and both integrals exist and are finite. To get the boundedness we simply use a standard inequality to estimate the integrals:

$$|\langle T, \varphi \rangle| \leq \int_{-\infty}^{-1} + \int_1^\infty \frac{|\varphi(x)|}{|x|} \, dx + \int_{-1}^1 |\varphi'(y(x))| \, dx.$$

If φ is supported in $|x| \leq R$ then

$$\int_{-\infty}^{-1} + \int_1^\infty \frac{|\varphi(x)|}{|x|} \, dx = \int_{-R}^{-1} + \int_1^R \frac{|\varphi(x)|}{|x|} \, dx \leq 2 \log R \, \|\varphi\|_\infty$$

and

$$\int_{-1}^1 |y'(y(x))| \, dx \leq 2 \|\varphi'\|_\infty.$$

Thus altogether

$$|\langle T, \varphi \rangle| \le 2 \log R \, ||\varphi||_\infty + 2||\varphi'||_\infty$$

which is an estimate of the required form with $m = 1$. Perhaps you are surprised that the estimate involves a derivative, while the original formula for T did not appear to involve any derivatives. But, of course, that derivative does show up in the existence proof.

Finally, you should beware confusing the continuity of a single distribution with the notion of limits of a sequence of distributions. In the first case we hold the distributions fixed and vary the test functions, while in the second case we hold the test function fixed and vary the distributions, because we defined $\lim_{j \to \infty} T_j = T$ to mean $\lim_{j \to \infty} \langle T_j, \varphi \rangle = \langle T, \varphi \rangle$ for every test function. This is actually a much simpler definition, but of course each distribution T_j and T must be continuous in the sense we have discussed.

6.6   Approximation by test functions

When we first discussed how to define various operations on distributions, we began with the important theorem that all distributions, no matter now rough or how rapidly growing, may be approximated by test functions, which are infinitely smooth and have compact support. How is this possible? In this section I will explain the procedure. As you will see, it involves two operations on distributions—convolution and multiplication by a smooth function—so that you might well raise the objection of circular reasoning: we used the theorem to define the operations, and now we use the operations to prove the theorem. In fact, to give a strictly logical exposition, you would have to use the method of adjoint identities to define the operations, then prove the theorem, and finally use the theorem to show that operations defined by the adjoint identities are the same as the operations defined by approximation by test functions. However, for this account, the important point to understand is the approximation procedure, because the same ideas are used in proving many other approximation theorems.

The two components of the approximation procedure are

1. multiplication by a smooth cut-off function, and

2. convolution with an approximate identity.

They can be performed in either order. Since we have already discussed (2), let's begin with (1), which is in many ways much simpler. A cut-off function is a test function whose graph looks like

It is identically one on a large set and then falls off to zero. Actually we want to think about a family of cut-off functions, with the property that

Figure 6.4

the set on which the function is one gets longer and larger. To be specific, let ψ be one fixed test function, say with support in $|x| \leq 2$, such that $\psi(x) = 1$ for $|x| \leq 1$. Then look at the family $\psi(\epsilon x)$ as $\epsilon \to 0$. We have $\psi(\epsilon x) = 1$ if $|\epsilon x| \leq 1$, or equivalently, if $|x| \leq \epsilon^{-1}$, and $\epsilon^{-1} \to \infty$ as $\epsilon \to 0$. Since $\psi(\epsilon x) \in \mathcal{D}$, we can form $\psi(\epsilon x)T$ for any distribution T, and it should come as no surprise that $\psi(\epsilon x)T \to T$ as $\epsilon \to 0$ as distributions. Indeed if φ is any test function, then φ has support in $|x| \leq R$ for some R, and then $\varphi(x)\psi(\epsilon x) = \varphi(x)$ if $\epsilon < 1/R$ (for then $\psi(\epsilon x) = 1$ on the support of φ). Thus

$$\langle \psi(\epsilon x)T, \varphi \rangle = \langle T, \psi(\epsilon x)\varphi \rangle = \langle T, \varphi \rangle$$

for $\epsilon < 1/R$, hence trivially $\lim_{\epsilon \to 0} \langle \psi(\epsilon x)T, \varphi \rangle = \langle T, \varphi \rangle$, which is what we mean by $\psi(\epsilon x)T \to T$ as $\epsilon \to 0$. But it is equally trivial to show that $\psi(\epsilon x)T$ is a distribution of compact support: in fact its support is contained in $|x| \leq 2/\epsilon$ because $\psi(\epsilon x) = 0$ if $|x| \geq 2/\epsilon$. To summarize: by the method of multiplication by cut-off functions, *every distribution may be approximated by distributions of compact support*. To show that every distribution can be approximated by test functions, it remains to show that every distribution of compact support can be approximated by test functions.

Now if T is a distribution of compact support and φ is a test function, then $\varphi * T$ is also a test function. In fact we have already seen that supp $\varphi * t \subseteq$ supp $\varphi +$ supp T (in the sense of sum sets) so $\varphi * T$ has compact support, and $\varphi * T$ is the function $\varphi * T(x) = \langle T, \tau_{-x}\tilde{\varphi} \rangle$ (recall that $\tilde{\varphi}(x) = \varphi(-x)$), which is C^∞ because all derivatives can be put on φ. Let $\varphi_\epsilon(x)$ be an approximate identity (so $\varphi_\epsilon(x) = \epsilon^{-n}\varphi(\epsilon^{-1}x)$ where $\int \varphi(x)\, dx = 1$). Then we claim $\lim_{\epsilon \to 0} \varphi_\epsilon * T = T$. Indeed, if ψ is any test function then $\langle \varphi_\epsilon * T, \psi \rangle = \langle T, \tilde{\varphi}_\epsilon * \psi \rangle$ and $\tilde{\varphi}_\epsilon$ is also an approximate identity, so $\tilde{\varphi}_\epsilon * \psi \to \psi$ hence $\langle T, \tilde{\varphi}_\epsilon * \psi \rangle \to \langle T, \psi \rangle$.

Actually, this argument requires that $\tilde{\varphi}_\epsilon * \psi \to \psi$ in $\mathcal{D}$, as described in the last section. That means we need to show that all the supports of $\tilde{\varphi}_\epsilon * \psi$ remain in a fixed compact set, and that all derivatives $(\partial/\partial x)^\alpha (\tilde{\varphi}_\epsilon * \psi)$ converge uniformly to $(\partial/\partial x)^\alpha \psi$. But both properties are easy to see, since supp $(\tilde{\varphi} * \psi) \subseteq$ supp $\tilde{\varphi}_\epsilon +$ supp ψ and supp $\tilde{\varphi}_\epsilon$ shrinks as $\epsilon \to 0$, while

$(\partial/\partial x)^{\alpha}(\tilde{\varphi}_{\epsilon} * \psi) = \tilde{\varphi}_{\epsilon} * (\partial/\partial x)^{\alpha}\psi$ and the approximate identity theorem shows that this converges uniformly to $(\partial/\partial x)^{\alpha}\psi$.

We can combine the two approximation processes into a single, two-step procedure. Starting with any distribution T, form $\varphi_{\epsilon} * (\psi(\epsilon x)T)$. This is a test function, and it converges to T as $\epsilon \to 0$ (to get a sequence, set $\epsilon = 1/k, k = 1, 2, \ldots$). It is also true that we can do it in the reverse order: $\psi(\epsilon x)(\varphi_{\epsilon} * T)$ is a test function (not the same one as before), and these also approximate T as $\epsilon \to 0$. The proof is similar, but slightly different.

The same procedure shows that functions in $\mathcal{D}$ can approximate functions in $\mathcal{S}$ or $\mathcal{E}$, and distributions in $\mathcal{S}'$ and $\mathcal{E}'$, where in each case the approximation is in the sense appropriate to the object being approximated. For example, if $f \in \mathcal{S}$ then we can regard f as a distribution, and so $\varphi_{\epsilon} * (\psi(\epsilon x)f(x)) \to f$ as a distribution: but we are asserting more, that the approximation is in terms of $\mathcal{S}$ convergence, i.e.,

$$(1 + |x|)^N \left(\frac{\partial}{\partial x}\right)^{\alpha} \varphi_{\epsilon} * (\psi(\epsilon x)f(x)) \to (1 + |x|)^N \left(\frac{\partial}{\partial x}\right)^{\alpha} f(x)$$

uniformly as $\epsilon \to 0$.

Now we can explain more clearly why we are allowed to think of tempered distributions as a subclass of distribution. Suppose T is a distribution (so $\langle T, \varphi \rangle$ is defined initially only for $\varphi \in \mathcal{D}$), and T satisfies and estimate of the form

$$|\langle T, \varphi \rangle| \leq c \sum_{|\alpha| \leq m} \left\| (1 + |x|)^N \left(\frac{\partial}{\partial x}\right)^{\alpha} \varphi(x) \right\|_{\infty}$$

for all $\varphi \in \mathcal{D}$. Then I claim there is a natural way to extend T so that $\langle T, \varphi \rangle$ is defined for all $\varphi \in \mathcal{S}$. Namely, we approximate $\varphi \in \mathcal{S}$ by test functions $\varphi_k \in \mathcal{D}$, in the above sense. Because of the estimate that T satisfies, it follows that $\langle T, \varphi_k \rangle$ converges as $k \to \infty$, and we define $\langle T, \varphi \rangle$ to be this limit. It is not hard to show that the limit is unique—it depends only on φ and not on the particular approximating sequence. And, of course, T satisfies the same estimate as before, but now for all $\varphi \in \mathcal{S}$. So the extended T is indeed continuous on $\mathcal{S}$, hence is a tempered distribution. The extension is unique, so we are justified in saying the original distribution *is* a tempered distribution, and the existence of an estimate of the given form is the condition that characterizes tempered distributions among all distributions. In practice it is usually not difficult to decide if a given distribution satisfies such an estimate, although there are some general existence theorems for distributions where it is either unknown, or very difficult, to decide. This is not just an academic question, since, as we have seen, Fourier transform methods require that the distribution be tempered.

6.7   Local theory of distributions

From the beginning, we have emphasized that distributions can be defined on any open set Ω in $\mathbb{R}^n$. Nevertheless, we have devoted most of our attention to distributions defined on all of $\mathbb{R}^n$. In this section we will consider more closely the distinction, and some of the subtler issues that arise.

Recall that the space of test functions $\mathcal{D}(\Omega)$ is defined to be the C^∞ functions φ with compact support in Ω. In other words, φ must be zero outside a compact set K which lies in the open set Ω, so K must have positive distance to the complement of Ω. Of course $\mathcal{D}(\Omega) \subseteq \mathcal{D}(\mathbb{R}^n)$. Then $\mathcal{D}'(\Omega)$, the distributions on Ω, are the continuous linear functionals on $\mathcal{D}(\Omega)$, where continuity means exactly the same thing as for distributions on $\mathbb{R}^n$, only restricted to test functions in $\mathcal{D}(\Omega)$. The two natural questions we need to answer are

1. what is the relationship between $\mathcal{D}'(\Omega)$ and $\mathcal{D}'(\mathbb{R}^n)$, and

2. what is the relationship between $\mathcal{D}'(\Omega)$ and the distributions of compact support whose support is contained in Ω?

Suppose $T \in \mathcal{D}'(\mathbb{R}^n)$. Then $\langle T, \varphi \rangle$ is defined for any $\varphi \in \mathcal{D}(\mathbb{R}^n)$, and since $\mathcal{D}(\Omega) \subseteq \mathcal{D}(\mathbb{R}^n)$ we obtain a distribution in $\mathcal{D}'(\Omega)$ by *restriction*. Let $\mathcal{R}T$ denote this restriction:

$$\langle \mathcal{R}T, \varphi \rangle = \langle T, \varphi \rangle \quad \text{for } \varphi \in \mathcal{D}(\Omega).$$

Although $\mathcal{R}T$ is defined by the same formula as T, it is different from T since it forgets about what T does outside Ω. In particular, $\mathcal{R}$ is not one-to-one because we can easily have $\mathcal{R}T_1 = \mathcal{R}T_2$ with $T_1 \neq T_2$ (just add a distribution supported outside Ω to T_1 to get T_2). Do we get all distributions in $\mathcal{D}'(\Omega)$ by restriction of distributions in $\mathcal{D}'(\mathbb{R}^n)$? In other words, is the mapping $\mathcal{R} : \mathcal{D}'(\mathbb{R}^n) \to \mathcal{D}'(\Omega)$ onto? Absolutely not! This is what makes the local theory of distributions interesting.

Let's give an example of a local distribution that is not the restriction of a global distribution. Let $\Omega = (0,1) \subseteq \mathbb{R}^1$. Consider $T \in \mathcal{D}'(\Omega)$ given by

$$\langle T, \varphi \rangle = \sum_{n=2}^{\infty} \varphi^{(n)}\left(\frac{1}{n}\right).$$

For any $\varphi \in \mathcal{D}(0,1)$ this is a finite sum, since φ must be zero in a neighborhood of 0, and it is easy to see that T is linear and continuous on $\mathcal{D}(0,1)$. But clearly T is not the restriction of any distribution in $\mathcal{D}'(\mathbb{R}^1)$ because the structure theorem would imply that such a distribution has locally finite order, which T manifestly does not. Another example is

$$\langle T, \varphi \rangle = \sum_{n=2}^{\infty} \varphi \left(\frac{1}{n} \right),$$

which is not a restriction since we can construct $\varphi \in \mathcal{D}(\mathbb{R}^1)$ supported in $[0, 1]$ for which the sum diverges.

What these examples show, and what is true in general, is that local distributions in $\mathcal{D}'(\Omega)$ can behave arbitrarily as you approach the boundary of Ω (because all the test functions in $\mathcal{D}(\Omega)$ vanish before you reach the boundary), but restrictions cannot have too rapid growth near the boundary. But now suppose T is a local distribution that is a restriction, $T = \mathcal{R}T_1$ for some $T_1 \in \mathcal{D}'(\mathbb{R}^n)$. Can we find an extension? We know that T does not uniquely define T_1, but is there some smallest or best extension? Could we, for example, find an extension T_1 that has support in $\bar{\Omega}$ (the closure of Ω)? Again the answer is no.

As an example, consider

$$\langle T, \varphi \rangle = \int_0^1 \frac{\varphi(x)}{x} \, dx$$

for $\varphi \in \mathcal{D}(0, 1)$. Since φ vanishes near zero the integral is finite (i.e., $1/x$ is locally integrable on $(0, 1)$). We have seen how to extend T to a distribution on $\mathbb{R}^1$, as for example

$$\langle T, \varphi \rangle = \lim_{\epsilon \to 0} \int_{|x| \geq \epsilon} \frac{\varphi(x)}{x} \, dx.$$

But if we try to make the support of the extension $[0, 1]$ we would have to take a definition

$$\int_0^1 \frac{\varphi(x)}{x} \, dx + \sum_{j=0}^{N} c_j \varphi^{(j)}(0).$$

The trouble is that the finite sum $\sum_{j=0}^{N} c_j \varphi^{(j)}(0)$ is always well defined, while if we take φ with $\varphi(0) \neq 0$ the integral will diverge.

On the other hand, it is always possible to find an extension whose support is a small neighborhood of Ω, provided of course that T is a restriction. Indeed, if $T = \mathcal{R}T_1$ then also $T = \mathcal{R}(\psi T_1)$ where ψ is any C^∞ function equal to 1 on Ω. We then have only to construct such ψ that vanishes outside a small neighborhood of Ω.

Turning to the second question, we write $\mathcal{E}'(\Omega)$ for the distributions of compact support whose support is contained in Ω. These are global distributions, but it is also natural to consider them as local distributions.

The point is that if $T_1 \neq T_2$ are distinct distributions in $\mathcal{E}'(\Omega)$, then also $\mathcal{R}T_1 \neq \mathcal{R}T_2$, so that we do not lose any information by identifying distributions in $\mathcal{E}'(\Omega) \subseteq \mathcal{D}'(\mathbb{R}^n)$ with their restrictions in $\mathcal{D}'(\Omega)$. Thus $\mathcal{E}'(\Omega) \subseteq \mathcal{D}'(\Omega)$. Actually $\mathcal{E}'(\Omega)$ is a smaller space than the space of restrictions, since the support of a distribution in $\mathcal{E}'(\Omega)$ must stay away from the boundary of Ω. Thus, for example,

$$\langle T, \varphi \rangle = \int_0^1 \varphi(x) \, dx$$

is a distribution in $\mathcal{E}'(\mathbb{R})$ which restricts to a distribution in $\mathcal{D}'(0,1)$, but it is not in $\mathcal{E}'(0,1)$ since its support is the closed interval $[0,1]$.

We can also regard $\mathcal{E}'(\Omega)$ as a space of continuous linear functions on $\mathcal{E}(\Omega)$, the space of all C^∞ functions on Ω (no restriction on the growth at the boundary). In summary, we have the containments

$$\mathcal{D}(\Omega) \subseteq \mathcal{E}'(\Omega) \subseteq \mathcal{R}\mathcal{D}'(\mathbb{R}^n) \subseteq \mathcal{D}'(\Omega)$$

and also $\mathcal{E}(\Omega) \subseteq \mathcal{D}'(\Omega)$, but no other containments are valid.

6.8 Distributions on spheres

The theory of distributions can be extended to smooth surfaces in $\mathbb{R}^n$ and, more generally, smooth manifolds. To keep the discussion on a concrete level, I will consider the case of the sphere $S^{n-1} \subseteq \mathbb{R}^n$ given by the equation $|x| = 1$. Note that $S^1 \subseteq \mathbb{R}^2$ is just the unit circle in the plane, and $S^2 \subseteq \mathbb{R}^3$ is the usual unit sphere in space. What we would like to get at is a theory that is intrinsic to the sphere itself, rather than the way it sits in Euclidean space. However, we will shamelessly exploit the presentation of S^{n-1} in $\mathbb{R}^n$ in order to simplify and clarify the discussion.

Before beginning, I should point out that we already have a theory of distributions connected with the sphere, namely, the distributions on $\mathbb{R}^n$ whose support is contained in S^{n-1}. For example, we have already considered the distribution

$$\langle T, \varphi \rangle = \int_{S^2} \varphi(x) \, d\sigma(x)$$

in connection with the wave equation. A simpler example is

$$\langle T, \varphi \rangle = \frac{\partial}{\partial x_1} \varphi(1, 0, \dots, 0).$$

However, this class of distributions really does depend on the embedding of S^{n-1} in $\mathbb{R}^{n-1}$, and so is extrinsic rather than intrinsic. Later we will see that it is a larger class than the distributions $\mathcal{D}'(S^{n-1})$ that we are going to define.

The idea is quite simple. In imitation of the definition of $\mathcal{D}'(\mathbb{R}^n)$, we want to define $\mathcal{D}'(S^{n-1})$ to be the continuous linear functionals on a space of test functions $\mathcal{D}(S^{n-1})$. These test functions should be just the C^∞ functions on the sphere (since the sphere is compact, we do not have to restrict the support of the test functions in any way). One way to define a C^∞ function on S^{n-1} is just as the restriction to the sphere of a C^∞ function on the ambient space $\mathbb{R}^n$; equivalently, φ defined on S^{n-1} is C^∞ if it can be extended to a C^∞ function on $\mathbb{R}^n$. Of course this definition appears to be extrinsic, and the extension is by no means unique (in fact there can also be many extensions that are not C^∞). To make matters better, it turns out that we can specify the extension. Let $\psi(r)$ be a C^∞ function of one variable whose graph looks like this

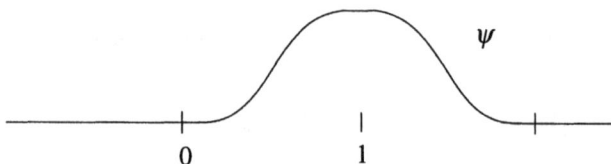

Figure 6.5

Then, given φ defined on S^{n-1}, let

$$E\varphi(x) = \psi(|x|)\varphi\left(\frac{x}{|x|}\right)$$

be the extension that multiplies the angular behavior of φ by the radial behavior of ψ (for any $x \neq 0$, the point $x/|x|$ belongs to S^{n-1}, so $\varphi\left(x/|x|\right)$ is well defined, and we set $E\varphi(0) = 0$ since $\psi(0) = 0$). Then φ has a C^∞ extension if and only if $E\varphi$ is C^∞. So we can define $\mathcal{D}(S^{n-1})$ as the space of functions φ on S^{n-1} such that $E\varphi \in \mathcal{D}(\mathbb{R}^n)$.

Still, we have not overcome the objection that this is an extrinsic definition. To get an equivalent intrinsic definition we need to consider the idea of local coordinates on the sphere. You are no doubt familiar with polar coordinates in $\mathbb{R}^2$ and spherical coordinates in $\mathbb{R}^3$:

$$x = r\cos\theta \qquad\qquad\qquad x = r\cos\theta\sin\phi$$
$$y = r\sin\theta \qquad\qquad\qquad y = r\sin\theta\sin\phi$$
$$z = r\cos\phi$$

We obtain points on the sphere by setting $r = 1$, which leaves us one (θ) and two (θ, ϕ) coordinates for S^1 and S^2. In other words, $(\cos\theta, \sin\theta)$ gives a point on S^1 for any choice of θ, and $(\cos\theta\sin\phi, \sin\theta\sin\phi, \cos\phi)$ gives a point on the sphere for any choice of θ, ϕ (you can think of θ as the longitude and $\frac{\pi}{2} - \phi$ as the latitude.) Of course we must impose some restrictions on the coordinates if we want a unique representation of each point, and here is where the locality of the coordinates system comes in.

For the circle, we could restrict θ to satisfy $0 \le \theta < 2\pi$ to have one θ value for each point. But including the endpoint 0 raises certain technical problems, so we prefer to consider a pair of local coordinate systems to cover the circle, say $0 < \theta < \frac{3\pi}{2}$ and $-\pi < \theta < \frac{\pi}{2}$, each given by an open interval in the parameter space and each wrapping $\frac{3}{4}$ of the way around the circle, with overlaps:

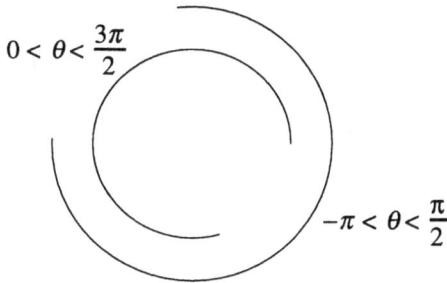

$$0 < \theta < \frac{3\pi}{2}$$

$$-\pi < \theta < \frac{\pi}{2}$$

Figure 6.6

Each local coordinate system describes a portion of the circle by a one-to-one smooth mapping from the coordinate θ to the circle, $\theta \to (\cos\theta, \sin\theta)$. In this case, the mapping formula is the same for both local coordinate systems, but this is a coincidence, so we do not want to emphasize the fact.

For the sphere, we can restrict θ to $0 \le \theta < 2\pi$ and ϕ to $0 \le \phi \le \pi$ to get a representation that is almost unique (when $\phi = 0$ or π, at the north and south poles, θ can assume any value—what time is it at the north pole?). But the situation here is even worse, because the north and south poles are singularities of the coordinate system (not of the sphere!). If we consider the local coordinate system given by spherical coordinates with say $0 < \theta < \frac{3\pi}{2}$ and $\frac{\pi}{6} < \phi < \frac{5\pi}{6}$, then we will get a smooth one-to-one parametrization of a portion of the sphere. For a second coordinate system we could take $-\pi < \theta < \frac{\pi}{2}$ and $\frac{\pi}{6} < \phi < \frac{5\pi}{6}$, and together they cover all of the sphere except for small circular caps around the north and south poles. We can cover these by similar rotated spherical coordinate systems (just interchange y and z, for example) centered about east and west poles. The point is that there are many such coordinate systems, and we can think of them as giving an atlas of local "maps" of portions of the sphere. These local coordinate systems are intrinsic to the sphere, and they give the sphere its structure as a two-dimensional manifold (the dimension is the number of coordinates).

Now we can state the intrinsic definition of a C^∞ function on the sphere: it is a function that is C^∞ as a function of the coordinates (the coordinates vary in an open set in a Euclidean space of the same dimension as the sphere) for every local coordinate system. In fact it is enough to verify this for a set of local coordinate system that covers the sphere, and then it is true for all local coordinate systems. It turns out that this intrinsic definition is equivalent to the extrinsic definition given before.

Once we have $\mathcal{D}(S^{n-1})$ defined, we can define $\mathcal{D}'(S^{n-1})$ as the continuous linear functionals on $\mathcal{D}(S^{n-1})$. Continuity requires that we define convergence in $\mathcal{D}(S^{n-1})$. From the extrinsic viewpoint, $\lim_{j\to\infty} \varphi_j = \varphi$ in $\mathcal{D}(S^{n-1})$ just means $\lim_{j\to\infty} E\varphi_j = E\varphi$ in $\mathcal{D}(\mathbb{R}^n)$. Intrinsically, it means φ_j converges uniformly to φ and all derivatives with respect to the coordinate variables also converge (a minor technicality arises in that the uniformity of convergence is only local—away from the boundary of the local coordinate system). Because the sphere is compact we do not have to impose the requirement of a common compact support.

Now if $T \in \mathcal{D}'(S^{n-1})$, we can also consider it as a distribution on $\mathbb{R}^n$, say $\widetilde{T}$, via the identity

$$\langle \widetilde{T}, \varphi \rangle = \langle T, \varphi|_{S^{n-1}} \rangle$$

where $\varphi|_{S^{n-1}}$ denotes the restriction of $\varphi \in \mathcal{D}(\mathbb{R}^n)$ to the sphere, and of course $\varphi|_{S^{n-1}} \in \mathcal{D}(S^{n-1})$. It comes as no surprise that $\widetilde{T}$ has support contained in S^{n-1}. But not every distribution supported on the sphere comes from a distribution in $\mathcal{D}'(S^{n-1})$ in this way. For example $\langle T, \varphi \rangle = (\partial/\partial x_1)\varphi(1, 0, \ldots, 0)$ involves a derivative in a direction perpendicular to the sphere and so cannot be computed in terms of the restriction $\varphi|_{S^{n-1}}$. In fact, we have the following structure theorem.

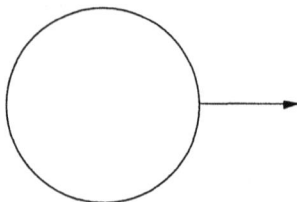

Figure 6.7

Theorem 6.8.1 *Every distribution $T \in \mathcal{D}'(\mathbb{R}^n)$ supported on the sphere S^{n-1} can be written*

$$T = \sum_{k=0}^{N} \left(\frac{\partial}{\partial r} \right)^k \tilde{T}_k$$

where $\tilde{T}_k$ is a distribution arising from $T_k \in \mathcal{D}'(S^{n-1})$ by the above identification, and

$$\frac{\partial}{\partial r} = \sum_{j=1}^{n} x_j \frac{\partial}{\partial x_j}$$

(in general we have to write

$$\frac{\partial}{\partial r} = \sum_{j=1}^{n} \frac{x_j}{|x|} \frac{\partial}{\partial x_j},$$

but $|x| = 1$ on the sphere, and of course $\partial/\partial r$ is undefined at the origin.)

The same idea can be used to define distributions $\mathcal{D}'(\mathcal{E})$ for any smooth surface $\sum \subseteq \mathbb{R}^n$. If $\sum$ is compact (e.g, on ellipsoid $\sum_{j=1}^{n} \lambda_j x_j^2 = 1, \lambda_j > 0$) there is virtually no change, while if $\sum$ is not compact (e.g, a hyperboloid $x_1^2 + x_2^2 - x_3^2 = 1$) it is necessary to insist that test functions in $\mathcal{D}(\sum)$ have compact support. (Readers who are familiar with the definition of an abstract C^∞ manifold can probably guess how to define test functions and distributions on a manifold.)

Here is one more interesting structure theorem involving $\mathcal{D}'(S^{n-1})$ for $n \geq 2$. A function that is homogeneous of degree α (remember this means $f(\lambda x) = \lambda^\alpha f(x)$ for all $\lambda > 0$) can be written $f(x) = |x|^\alpha F(x/|x|)$ where F is a function on the sphere. Could the same be true for homogeneous distributions? It turns out that it is true if $\alpha > -n$ (or even complex α with $\mathrm{Re}\,\alpha > -n$). Let $T \in \mathcal{D}'(S^{n-1})$. We want to make sense of the symbol $|x|^\alpha T (x/|x|)$. If T were a function on the sphere we could use polar coordinates to write

$$\int |x|^\alpha T \left(\frac{x}{|x|} \right) \varphi(x)dx = \int_0^\infty \left(\int_{S^{n+1}} T(y)\varphi(ry)\, d\sigma(y) \right) r^{\alpha+n-1} dr$$

for $\varphi \in \mathcal{D}(\mathbb{R}^n)$. Thus $|x|^\alpha T (x/|x|)$ should be given in the general case by $\int_0^\infty \langle T, \varphi(ry) \rangle r^{\alpha+n-1} dr$.

Theorem 6.8.2 *If $\mathrm{Re}\,\alpha > -n$ then this expression defines a distribution on $\mathbb{R}^n$ that is homogeneous of degree α, for any $T \in \mathcal{D}'(S^{n-1})$. Conversely, every distribution on $\mathbb{R}^n$ that is homogeneous of degree α is of this form, for some $T \in \mathcal{D}'(S^{n-1})$.*

The situation when Re $\alpha \le -n$ is more complicated, because the r-integral may be divergent. Nevertheless, there is a way of interpreting the integral that makes the analog of the theorem true for all α except $\alpha = -n, -n-1, \ldots$ To see what happens at the exceptional points, let's look at $\alpha = -n$. Then the integral is

$$\int_0^\infty \langle T, \varphi(ry) \rangle \frac{dr}{r}$$

which will certainly diverge if $\langle T, \varphi(ry) \rangle$ does not tend to zero as $r \to 0$. But if $T \in \mathcal{D}'(S^{n-1})$ has the property that $\langle T, 1 \rangle = 0$ (1 is the constant function on S^{n-1}, which is a test function) then $\langle T, \varphi(ry) \rangle = \langle T, \varphi(ry) - \varphi(0)1 \rangle$ and since $\varphi(ry) - \varphi(0)1 \to 0$ as $r \to 0$ (in the sense of $\mathcal{D}(S^{n-1})$ convergence) we do have $\langle T, \varphi(ry) \rangle \to 0$ as $r \to 0$, and the integral converges. Thus the theorem generalizes to say $\int_0^\infty \langle T, \varphi(ry) \rangle dr/r$ is a distribution homogeneous of degree $-n$ provided $\langle T, 1 \rangle = 0$. However, the converse says that every homogeneous distribution of degree $-n$ can be written as a sum of such a distribution and a multiple of the delta distribution.

In a sense, there is a perfect balance in this result: we pick up a one-dimensional space, the multiples of δ, which have nothing to do with the sphere; at the same time we give up a one-dimensional space on the sphere by imposing the condition $\langle T, 1 \rangle = 0$. An analogous balancing happens at $\alpha = -n - k$, but it involves spaces of higher dimension.

6.9 Problems

1. Show that supp $\varphi' \subseteq$ supp φ for every test function φ on $\mathbb{R}^1$. Give an example where supp $\varphi' \ne$ supp φ.

2. Show that supp $\left(\dfrac{\partial}{\partial x}\right)^\alpha T \subseteq$ supp (T) for every distribution T.

3. Show that supp $(fT) \subseteq$ supp T for every C^∞ function f.

4. Show that $\int_\Omega |f(x)|\,dx = 0$ if and only if both $\int_\Omega |f^+(x)|\,dx = 0$ and $\int_\Omega |f^-(x)|\,dx = 0$.

5. Suppose T is a distribution on $\mathbb{R}^1$, and define a distribution S on $\mathbb{R}^2$ by $\langle S, \varphi \rangle = \langle T, \varphi_0 \rangle$ where $\varphi_0(x) = \varphi(x, 0)$. Show that supp S is contained in the x-axis. Give an example of a distribution whose support is contained in the x-axis which is not of this form. (*Hint:* $(\partial/\partial y)\delta$.)

6. Let f_1 and f_2 be continuous functions on $\mathbb{R}^2$. Show that there exists a continuous function f such that

$$\frac{\partial f_1}{\partial x_1} + \frac{\partial f_2}{\partial x_2} = \frac{\partial^2 f}{\partial x_1 \partial x_2}.$$

7. Let $\langle T, \varphi \rangle = \lim_{\epsilon \to 0} \int_{-\infty}^{-\epsilon} + \int_{\epsilon}^{\infty} [\varphi(x)/x]\, dx$ on $\mathbb{R}^1$. Show that

$$T = \frac{d}{dx} \log |x| = \frac{d^2}{dx^2} (x \log |x|).$$

8. Give an example of a distribution T on $\mathbb{R}^1$ that does not have compact support but for which T' does have compact support.

9. Define an "integral" IT of a distribution T on $\mathbb{R}^1$ as follows: $\langle IT, \varphi \rangle = -\langle T, I\varphi \rangle$ where $I\varphi(x) = \int_{-\infty}^{x} \varphi(y)\, dy$ provided $\int_{-\infty}^{\infty} \varphi(y)\, dy = 0$ (this condition is needed to have $I\varphi \in \mathcal{D}$), and more generally $\langle IT, \varphi + a\varphi_0 \rangle = -\langle T, I\varphi \rangle$ where φ is as before and φ_0 is any fixed test function with $\int_{-\infty}^{\infty} \varphi(y)\, dy = 1$. Show that every test function is of this form and that IT is a well defined distribution. Show that $(d/dx)IT = T$. Show that any other choice of φ_0 yields a distribution differing by a constant from IT.

10. Let ψ_j be any sequence of nonnegative test functions such that $\sum \psi_j(x)$ is everywhere positive. Show that $\varphi_j(x) = \psi_j(x)/\sum_k \psi_k(x)$ gives a partition of unity.

11. Let $f(x) = \sin e^x$. Show that $f'(x)$ is a tempered distribution, even though we do not have $|f'(x)| \le c(1 + |x|)^N$ for any c and N.

12. Give the form of the general distribution whose support is the single point y.

13. Give the form of the general distribution whose support is a pair of points y and z.

14. Show that there exists a test function φ whose Taylor expansion can be prescribed arbitrarily at two distinct points y and z.

15. Show that the real and imaginary parts of $(x + iy)^k$ are harmonic polynomials on $\mathbb{R}^2$, for any positive integer k.

16. Show that if $fT = 0$ for a C^∞ function f, then $\operatorname{supp} T$ is contained in the zero-set of f (i.e., $\{x : f(x) = 0\}$). (*Hint*: the zero-set is closed.)

17. Show that the distribution

$$\langle T, \varphi \rangle = \int_{-\infty}^{-1} + \int_{1}^{\infty} \frac{\varphi(x)}{|x|} \, dx + \int_{-1}^{1} \frac{\varphi(x) - \varphi(0)}{|x|} \, dx$$

is not positive. (*Hint*: choose φ with a maximum at $x = 0$.)

18. Show that $\operatorname{supp} f = \operatorname{supp} f^{+} \cup \operatorname{supp} f^{-}$ for continuous f. Does the same hold even if f is not continuous?

19. Show that $\langle T, \varphi \rangle \geq 0$ if T is a positive distribution of compact support and $\varphi \geq 0$ on a neighborhood of the support of T.

20. Give an example of a test function φ such that φ^{+} and φ^{-} are not test functions.

21. Show that any real-valued test function φ can be written $\varphi = \varphi_1 - \varphi_2$ where φ_1 and φ_2 are nonnegative test functions. Why does not this contradict problem 20?

22. Define the Cantor function f to be the continuous, nondecreasing function that is zero on $(-\infty, 0]$, one on $[1, \infty)$, and on the unit interval is constant on each of the deleted thirds in the construction of the Cantor set, filling in half-way between

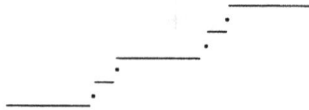

Figure 6.8

$\left(\text{so } f(x) = \frac{1}{2} \text{ on the first stage middle third } \left[\frac{1}{3}, \frac{2}{3}\right], f(x) = \frac{1}{4} \text{ and } \frac{3}{4} \text{ on the second stage middle thirds, and so on}\right)$. Show that the Cantor measure is equal to the distributional derivative of the Cantor function.

23. Show that the total length of the deleted intervals in the construction of the Cantor set is one. In this sense the total length of the Cantor set is zero.

24. Give an example of a function on the line for which $\|\varphi\|_{\infty} \leq 10^{-6}$, $\|\varphi'\|_{\infty} \leq 10^{-6}$ but $\|\varphi''\|_{\infty} \geq 10^{6}$. (*Hint*: Try $a \sin bx$ for suitable constants a and b.)

25. Let φ be a nonzero test function on $\mathbb{R}^1$. Does the sequence $\varphi_n(x) = \frac{1}{n}\varphi(x-n)$ tend to zero in the sense of $\mathcal{D}$? Can you find an example of a distribution T for which $\lim_{n\to\infty}\langle T, \varphi_n\rangle \neq 0$? Can you find a tempered distribution T with $\lim_{n\to\infty}\langle T, \varphi_n\rangle \neq 0$?

26. Let T be a distrbution and φ a test function. Show that

$$(\partial/\partial x_j)(T * \varphi) = T * (\partial\varphi/\partial x_j)$$

by bringing the derivative inside the distribution.

27. For Re $\alpha < 2, \alpha \neq 1$, define a distribution T_α concocted from $|x|^{-\alpha}$ by

$$\langle T_\alpha, \varphi\rangle = \int_{-\infty}^{-1} + \int_1^\infty \varphi(x)|x|^{-\alpha}dx$$
$$+ \int_{-1}^1 (\varphi(x) - \varphi(0))|x|^{-\alpha}dx + \frac{2\varphi(0)}{1-\alpha}.$$

Show that $\langle T_\alpha, \varphi\rangle = \int_{-\infty}^\infty \varphi(x)|x|^{-\alpha}dx$ if $\varphi(0) = 0$ or if Re $\alpha < 1$. Show that T is defined for all test functions and satisfies a boundedness condition with $m = 1$.

28. Show that T_α is homogeneous of degree $-\alpha$. Would the same be true if we omitted the term $2\varphi(0)/(1-\alpha)$?

29. Show that the family of distributions T_α is analytic as a function of α (i.e., $\langle T_\alpha, \varphi\rangle$ is analytic for every test function φ) in Re $\alpha < 2, \alpha \neq -1$.

30. Show that if T is a distribution and f a C^∞ function, the definition $\langle fT, \varphi\rangle = \langle T, f\varphi\rangle$ yields a distribution (in the sense that fT satisfies a boundedness condition).

31. Let $\langle T, \varphi\rangle = \int f(x)\varphi(x)\, dx$ for a function $f(x)$ satisfying $|f(x)| \leq c(1 + |x|)^N$. What kind of boundedness condition as a tempered distribution does T satisfy?

32. Let T be a distribution of compact support and f a polynomial. Show that $T * f$ is also a polynomial. (*Hint*: Polynomials can be characterized by the condition $(\partial/\partial x)^\alpha f = 0$ for some α.) What can you say about the degree of the polynomial $T * f$?

33. Show that $\langle T, \varphi\rangle = \sum_{n=1}^\infty \varphi(n)$ is a tempered distribution on $\mathbb{R}^1$.

34. Show that every distribution in $\mathcal{D}'(\Omega)$ can be approximated by test functions in $\mathcal{D}(\Omega)$ (you will have to use the fact that there exists a sequence ψ_j of test functions in $\mathcal{D}(\Omega)$ such that $\psi_j = 1$ on a set K_j with $\Omega = \cup K_j$).

35. For $\Omega = (0,1) \subseteq \mathbb{R}^1$ construct ψ_j as called for in problem 34.

36. Suppose $\Omega_1 \subseteq \Omega_2$. Show that distributions in $\mathcal{D}'(\Omega_2)$ can be restricted to distributions in $\mathcal{D}'(\Omega_1)$.

37. Give an example of a function in $\mathcal{E}(0,1)$ that is not the restriction of any function in $\mathcal{E}(\mathbb{R}^1)$.

38. Show that the product φT is well defined for $T \in \mathcal{D}'(\Omega)$ and $\varphi \in \mathcal{E}(\Omega)$.

39. Show that differentiation is well defined in $\mathcal{D}'(\Omega)$.

40. Show that distributions on the circle can be identified with periodic distributions on the line with period 2π.

41. Show that if $T \in \mathcal{D}'(S^{n-1})$ then

$$\left(x_j \frac{\partial}{\partial x_k} - x_k \frac{\partial}{\partial x_j} \right) T$$

is a well defined distribution in $\mathcal{D}'(S^{n-1})$. Show the same is true for $\mathcal{D}'(|x| < 1)$.

42. Compute the dimension of the space of distributions supported at the origin that are homogeous of degree $-n - k$.

Chapter 7

Fourier Analysis

7.1 The Riemann-Lebesgue lemma

One of the basic estimates for the Fourier transform we have used is

$$|\hat{f}(\xi)| \leq \int |f(x)| \, dx.$$

It shows that $\hat{f}$ is bounded if $\int |f(x)| \, dx$ is finite. But it turns out that something more is true: $\hat{f}(\xi)$ goes to zero as $\xi \to \infty$. What do we mean by this? In one dimension it means $\lim_{\xi \to +\infty} \hat{f}(\xi) = 0$ and $\lim_{\xi \to -\infty} \hat{f}(\xi) = 0$. In more than one dimension it means something slightly stronger than the limit of $\hat{f}(\xi)$ is zero along every curve tending to infinity.

Definition 7.1.1 *We say a function $g(\xi)$ defined on $\mathbb{R}^n$ vanishes at infinity, written $\lim_{\xi \to \infty} g(\xi) = 0$, if for every $\epsilon > 0$ there exists N such that $|g(\xi)| \leq \epsilon$ for all $|\xi| \geq N$.*

In other words, g is uniformly small outside a bounded set.

For a continuous function, vanishing at infinity is a stronger condition than boundedness (if f is continuous then $|f(x)| \leq M$ on $|x| \leq N$, and $|f(x)| \leq \epsilon$ on $|x| \geq N$). The Riemann-Lebesgue lemma asserts that $\hat{f}$ vanishes at infinity if $\int |f(x)| \, dx$ is finite. Riemann proved this for integrals in the sense of Riemann, and Lebesgue extended the result for his more general theory of integration. Whatever theory of integration you have learned, you can interpret the result in that theory. We will say that f is *integrable* if $\int |f(x)| \, dx$ is finite.

Let me sketch two different proofs, omitting certain details that involve integration theory. The most straightforward proof combines three ideas:

113

1. We already know the vanishing at infinity for $\hat{f}$ if $f \in S$, since this implies $\hat{f} \in S$.

2. S is dense in the integrable functions.

3. The property of vanishing at infinity is preserved under uniform limits.

The density of S (or even $\mathcal{D}$) in various distribution spaces was discussed in 6.6. Essentially the same construction, multiplication by a cut-off function and convolution with an approximate identity, yields the existence of a sequence of functions $f_n \in S$ such that

$$\lim_{n \to \infty} \int |f_n(x) - f(x)| \, dx = 0$$

for any integrable function f. This is the meaning of (2) above, and we will not go into the details because it involves integration theory. Notice that by using our basic estimate we may conclude that $\hat{f}_n$ converges uniformly to $\hat{f}$,

$$\lim_{n \to \infty} \sup_{\xi} |\hat{f}_n(\xi) - \hat{f}(\xi)| = 0.$$

Thus to conclude the proof of the Riemann-Lebesgue lemma we need to show:

Lemma 7.1.2 *If g_n vanishes at infinity and $g_n \to g$ uniformly, then g vanishes at infinity.*

The proof of this lemma is easy. Given the error $\epsilon > 0$, we break it in half and find g_n such that $|g_n(\xi) - g(\xi)| \leq \frac{\epsilon}{2}$ for all ξ (this is the uniform convergence). Then for that particular g_n, using the fact that it vanishes at infinity, we can find N such that $|g_n(\xi)| \leq \frac{\epsilon}{2}$ for $|\xi| \geq N$. But then for $|\xi| \geq N$ we have

$$
\begin{aligned}
|g(\xi)| &\leq |g_n(\xi) - g(\xi)| + |g_n(\xi)| \\
&\leq \frac{\epsilon}{2} + \frac{\epsilon}{2} = \epsilon
\end{aligned}
$$

so g vanishes at infinity.

This first proof is conceptually clear, but it uses the machinery of the Schwartz class S. Riemann did not know about S, so instead he used a clever trick. He observed that you can make the change of variable $x \to x + \pi/\xi$ in the definition of the Fourier transform (this is in one-dimension, but a simple variant works in higher dimensions) to obtain

$$
\begin{aligned}
\hat{f}(\xi) &= \int f(x)e^{ix\xi}\,dx \\
&= \int f(x+\pi/\xi)e^{ix\xi}e^{i\pi}\,dx \\
&= -\int f(x+\pi/\xi)e^{ix\xi}\,dx.
\end{aligned}
$$

So what? Well not much, until you get the even cleverer idea to average this with the original definition to obtain

$$
\hat{f}(\xi) = \frac{1}{2}\int (f(x) - f(x+\pi/\xi))e^{ix\xi}\,dx.
$$

Now take absolute values to estimate

$$
|\hat{f}(\xi)| \le \frac{1}{2}\int |f(x) - f(x+\pi/\xi)|\,dx.
$$

What happens when $\xi \to \infty$? Clearly $\pi/\xi \to 0$, and it is not surprising that

$$
\int |f(x) - f(x+\pi/\xi)|\,dx \to 0
$$

(this again requires some integration theory).

The Riemann-Lebesgue lemma is also true for Fourier sine and cosine transforms, namely

$$
\lim_{\xi\to\infty} \int f(x)\sin x \cdot \xi\,dx = 0
$$

and

$$
\lim_{\xi\to\infty} \int f(x)\cos x \cdot \xi\,dx = 0
$$

if f is integrable. In fact, these forms follow by considering $\hat{f}(\xi) \pm \hat{f}(-\xi)$.

As an application of the Riemann-Lebesgue lemma we prove the equipartition of energy for solutions of the wave equation. Remember that in section 5.3 we showed that the energy

$$
E(t) = \frac{1}{2}\int |u_t(x,t)|^2\,dx + \frac{1}{2}k^2\sum_{j=1}^{n}\int |u_{x_j}(x,t)|^2\,dx
$$

was a conserved quantity for solutions of the wave equation $u_{tt} - k^2\Delta_x u = 0$, where the first integral is interpreted as kinetic energy and the second as potential energy. The equipartition says that as $t \to \pm\infty$, the kinetic

and potential energies each tend to be half the total energy. To see this
we repeat the argument we used to prove the conservation of energy, but
retain just one of the two terms, say kinetic energy. This is equal to

$$\frac{1}{2(2\pi)^n} \int \left| \frac{\partial}{\partial t} \mathcal{F}_x u(\xi, t) \right|^2 d\xi$$

by the Plancherel formula, and using the formula

$$\mathcal{F}_x u(\xi, t) = \hat{f}(\xi) \cos kt|\xi| + \hat{g}(\xi) \frac{\sin kt|\xi|}{k|\xi|}$$

for the solution to the wave equation we compute

$$
\begin{aligned}
\left| \frac{\partial}{\partial t} \mathcal{F}_x u(\xi, t) \right|^2 &= k^2 |\xi|^2 \sin^2 kt|\xi| \, |\hat{f}(\xi)|^2 \\
&\quad + \cos^2 kt|\xi| \, |\hat{g}(\xi)|^2 \\
&\quad - k|\xi| \sin kt|\xi| \cos kt|\xi| \left(\hat{f}(\xi)\overline{\hat{g}(\xi)} + \overline{\hat{f}(\xi)}\hat{g}(\xi) \right) \\
&= \frac{1}{2} k^2 |\xi|^2 |\hat{f}(\xi)|^2 - \frac{1}{2} k^2 |\xi|^2 \cos 2kt|\xi| \, |\hat{f}(\xi)|^2 \\
&\quad + \frac{1}{2} |\hat{g}(\xi)|^2 + \frac{1}{2} \cos 2kt|\xi| \, |\hat{g}(\xi)|^2 \\
&\quad - \frac{1}{2} k|\xi| \sin 2kt|\xi| \left(\hat{f}(\xi)\overline{\hat{g}(\xi)} + \overline{\hat{f}(\xi)}\hat{g}(\xi) \right)
\end{aligned}
$$

using the familiar trigonometric identities for double angles. The point is
that when we take the ξ-integral we find that the kinetic energy is equal to
the sum of

$$\frac{1}{4} \frac{1}{(2\pi)^n} \int (|\hat{g}(\xi)|^2 + k^2 |\xi|^2 |\hat{f}(\xi)|^2) \, d\xi$$

which is exactly half the total energy, plus the terms

$$
\begin{aligned}
\frac{1}{4} \frac{1}{(2\pi)^n} \Bigg[&\int \cos 2kt|\xi| \, |\hat{g}(\xi)|^2 \, d\xi \\
&- k^2 \int \cos 2kt|\xi| \, \left| |\xi| \hat{f}(\xi) \right|^2 d\xi \\
&- k \int \sin 2kt|\xi| \left(|\xi| \hat{f}(\xi)\overline{\hat{g}(\xi)} + |\xi| \overline{\hat{f}(\xi)}\hat{g}(\xi) \right) d\xi \Bigg]
\end{aligned}
$$

which we claim must go to zero as $t \to \pm\infty$ by the Riemann-Lebesgue
lemma.

This argument is perhaps simpler to understand if we consider the case $n = 1$ first. Consider the term

$$\int \cos 2kt|\xi| \; |\hat{g}(\xi)|^2 \, d\xi.$$

We may drop the absolute values of $|\xi|$ because cosine is even. Now the requirement that the initial energy is finite implies $\int |\hat{g}(\xi)|^2 \, d\xi$ is finite, hence $|\hat{g}(\xi)|^2$ is integrable. Then $\lim_{t\to\pm\infty} \int \cos 2kt\xi |\hat{g}(\xi)|^2 \, d\xi = 0$ is the Riemann-Lebesgue lemma for the Fourier cosine transform of $|\hat{g}(\xi)|^2$. Similarly, we must have $|\xi \hat{f}(\xi)|^2$ integrable, hence the vanishing of $\int \cos 2kt\xi |\xi \hat{f}(\xi)|^2 \, d\xi$ as $t \to \pm\infty$. Finally, the function $\xi \hat{f}(\xi)\overline{\hat{g}(\xi)}$ is also integrable (this follows from the Cauchy-Schwartz inequality for integrals,

$$\int |F(x)G(x)| \, dx \leq \left(\int |F(x)|^2 \, dx \right)^{1/2} \left(\int |G(x)|^2 \, dx \right)^{1/2},$$

which is one of the standard estimates of integration theory), and so

$$\int \sin 2kt|\xi| \left(|\xi| \hat{f}(\xi)\overline{\hat{g}(\xi)} \right) d\xi = \int \sin 2kt\xi \left(\xi \hat{f}(\xi)\overline{\hat{g}(\xi)} \right) d\xi$$

vanishes as $t \to \pm\infty$.

The argument in $\mathbb{R}^n$ is similar. In fact, we appeal to the one-dimensional Riemann-Lebesgue lemma! The idea is to do the integration in polar coordinates, so that, for example,

$$\int_{\mathbb{R}^n} \cos 2kt|\xi| \; |\hat{g}(\xi)|^2 \, d\xi = \int_0^\infty \cos 2ktr \left(\int_{S^{n-1}} |\hat{g}(r\omega)|^2 \, d\omega \right) r^{n-1} \, dr.$$

The fact that $\int |\hat{g}(\xi)|^2 \, d\xi$ is finite is equivalent to $r^{n-1} \int_{S^{n-1}} |\hat{g}(r\omega)|^2 d\omega$ being integrable as a function of the single variable r, and then the one-dimensional Riemann-Lebesgue lemma gives

$$\lim_{t\to\pm\infty} \int_{\mathbb{R}^n} \cos 2kt|\xi| \; |\hat{g}(\xi)|^2 \, d\xi = 0.$$

The other terms follow by the same sort of reasoning.

The Riemann-Lebesgue lemma is a qualitative result: it tells you that $\hat{f}$ vanishes at infinity, but it does not specify the rate of decay. For many purposes it is desirable to have a quantitative estimate, such as $|\hat{f}(\xi)| \leq c|\xi|^{-\alpha}$ for $|\xi| \geq 1$ for some constants c and α, or some other comparison $|\hat{f}(\xi)| \leq ch(\xi)$ where h is an explicit function that vanishes at infinity. It turns out, however, that we cannot conclude that any such estimate holds from the mere fact that f is integrable; in fact, for any fixed function h vanishing at infinity, it is possible to find f which is continuous and has compact support, such that $|\hat{f}(\xi)| \leq ch(\xi)$ for all ξ does not hold for any constant c (this is the same as saying the ratio $\hat{f}(\xi)/h(\xi)$ is unbounded). This negative result is rather difficult to prove, but it is easy to give examples of integrable functions that do not satisfy $|\hat{f}(\xi)| \leq c|\xi|^{-\alpha}$ for all $|\xi| \geq 1$ for any constant c, where $\alpha > 0$ is specified in advance. Recall that we showed the Fourier transform of the distribution $|x|^{-\beta}$ was equal to a multiple of $|\xi|^{\beta-n}$, for $0 < \beta < n$ (the constant depends on β, but for this argument it is immaterial). Now $|x|^{-\beta}$ is never integrable, because we need $\beta < n$ for the finiteness of the integral near zero and $\beta > n$ for the finiteness of the integral near infinity. Since it is the singularity near zero that interests us, we form the function $f(x) = |x|^{-\beta}e^{-|x|^2/2}$. Multiplying by the Gaussian factor makes the integral finite at infinity, regardless of β. Thus our function is integrable as long as $\beta < -n$. Now the Fourier transform is a multiple of the convolution of $|\xi|^{\beta-n}$ and $e^{-|\xi|^2/2}$. Since these functions are positive, it is easy to show that their convolution decays exactly at the rate $|\xi|^{\beta-n}$. In fact we have the estimate

$$\begin{aligned}
\hat{f}(\xi) &= c \int |\xi - \eta|^{\beta-n} e^{-|\eta|^2/2} d\eta \\
&\geq c \int_{|\eta| \leq 1} |\xi - \eta|^{\beta-n} e^{-|\eta|^2/2} d\eta \\
&\geq c \int_{|\eta| \leq 1} |\xi - \eta|^{\beta-n} e^{-1/2} d\eta \\
&\geq c'(|\xi| - 1)^{\beta-n} \quad \text{if} \quad |\xi| \geq 2.
\end{aligned}$$

Once we take β to satisfy $\alpha - n < \beta < n$ we have our counterexample to the estimate

$$|\hat{f}(\xi)| \leq c|\xi|^{-\alpha} \quad \text{for} \quad |\xi| \geq 1.$$

   These examples (and the general negative result) show that the Riemann-Lebesgue lemma is *best possible*, or *sharp*, meaning that it cannot be improved, at least in the direction we have considered. Of course all such claims must be taken with a grain of salt, because there are always other possible directions in which a result can be improved. It is also important to realize that it is easy to put conditions on f that will imply specific decay rates on the Fourier transform; however, as far as we know now, all such conditions involve some sort of smoothness on f, such as differentiability or Hölder or Lipschitz conditions. The point of the Riemann-Lebesgue lemma is that we are only assuming a condition on the size of the function.

Another theorem that relates the size of a function and its Fourier transform is the Hausdorff-Young inequality, which says

$$\left(\int |\hat{f}(\xi)|^q \, d\xi \right)^{1/q} \leq c_p \left(\int |f(x)|^p \, dx \right)^{1/p}$$

whenever $|f|^p$ is integrable, where p and q are fixed numbers satisfying $\frac{1}{p} + \frac{1}{q} = 1$, and also $1 < p \leq 2$. In fact it is even known what the best constant c_p is (this is Beckner's inequality, and you compute the rather complicated formula for c_p by taking $f(x)$ to be a Gaussian and make the inequality an equality). Note that a special case of the Hausdorff-Young inequality, $p = q = 2$, is a consequence of the Plancherel formula; in fact in this case we have an equality rather than an inequality. Also, the estimate $|\hat{f}(\xi)| \leq \int |f(x)| \, dx$ is the limiting case of the Hausdorff-Young inequality as $p \to 1$. The proof of the Hausdorff-Young inequality is obtained from these two endpoint results by a method known as "interpolation," which is beyond the scope of this book.

7.2   Paley-Wiener theorems

We have seen that there is a relationship between decay at infinity for a function (or tempered distribution) and smoothness of the Fourier transform. We may think of compact support as the ultimate in decay at infinity, and analyticity as the ultimate in smoothness. Then it should come as no surprise that a function (or distribution) of compact support has a Fourier transform that is analytic. We use the term "Paley-Wiener theorem" to refer to any result that characterizes support information about f in terms of analyticity conditions on $\hat{f}$ (note that some Paley-Wiener theorems concern supports that are not compact). Generally speaking, it is rather easy to pass from support information about f to analyticity conditions on $\hat{f}$, but it is rather tricky to go in the reverse direction.

You can get the general idea rather quickly by considering the simplest example. Let f be a function on $\mathbb{R}^1$, and let's ask if it is possible to extend the definition of $\hat{f}$ from $\mathbb{R}$ to $\mathbb{C}$ so as to obtain a complex analytic function. The obvious idea is to replace the real variable ξ with the complex variable $\zeta = \xi + i\eta$ and substitute this in the definition:

$$
\begin{aligned}
\hat{f}(\xi + i\eta) &= \int_{-\infty}^{\infty} e^{ix(\xi + i\eta)} f(x) \, dx \\
&= \int_{-\infty}^{\infty} e^{ix\xi} e^{-x\eta} f(x) \, dx \\
&= \mathcal{F}_x(e^{-x\eta} f(x))(\xi).
\end{aligned}
$$

We see that this is just the ordinary Fourier transform of the function $e^{-x\eta}f(x)$. The trouble is, if we are not careful, this may not be well defined because $e^{-x\eta}$ grows too rapidly at infinity. But there is no problem if f has compact support, for $e^{-x\eta}$ is bounded on this compact set. Then it is easy to see that $\hat{f}(\zeta)$ is analytic, either by verifying the Cauchy-Riemann equations or by computing the complex derivative

$$\left(\frac{d}{d\zeta}\right)\hat{f}(\zeta) = \int_{-\infty}^{\infty} ixe^{ix\zeta}f(x)\,dx$$
$$= \mathcal{F}_x(ixe^{-x\eta}f(x))(\xi).$$

In fact, the same reasoning shows that if f is a distribution of compact support, we can define $\hat{f}(\zeta) = \langle f, e^{ix\zeta}\rangle$ and this is analytic with derivative $(d/d\zeta)\hat{f}(\zeta) = \langle f, ixe^{ix\zeta}\rangle$.

What kind of analytic functions do we obtain? Observe that $\hat{f}(\zeta)$ is an entire analytic function (defined and analytic on the entire complex plane). But not every entire analytic function can be obtained in this way, because $\hat{f}(\zeta)$ satisfies some growth conditions. Say, to be more precise, that f is bounded and supported on $|x| \leq A$. Then

$$|\hat{f}(\zeta)| = \left|\int_{-A}^{A} f(x)e^{ix\xi}e^{-x\eta}\,dx\right| \leq e^{A|\eta|}\int_{-A}^{A}|f(x)|\,dx = ce^{A|\eta|}$$

because $e^{-x\eta}$ achieves its maximum value $e^{A|\eta|}$ at one of the endpoints $x = \pm A$. Thus the growth of $|\hat{f}(\zeta)|$ is at most exponential in the imaginary part of ζ. Such functions are called entire functions of *exponential type* (or more precisely *exponential type A*). Note that it is not really necessary to assume that f is bounded—we could get away with the weaker assumption that f is integrable.

We can now state two Paley-Wiener theorems.

Theorem 7.2.1 *P.W. 1: Let f be supported in $[-A, A]$ and satisfy $\int_{-A}^{A}|f(x)|^2\,dx < \infty$. Then $\hat{f}(\zeta)$ is an entire function of exponential type A, and $\int_{-\infty}^{\infty}|\hat{f}(\xi)|^2\,d\xi < \infty$. Conversely, if $F(\zeta)$ is an entire function of exponential type A, and $\int_{-\infty}^{\infty}|F(\xi)|^2\,d\xi < \infty$, then $F = \hat{f}$ for some such function f.*

Theorem 7.2.2 *P.W. 2: Let f be a C^{∞} function supported in $[-A, A]$. Then $\hat{f}(\zeta)$ is an entire function of exponential type A, and $\hat{f}(\xi)$ is rapidly decreasing, i.e., $|\hat{f}(\xi)| \leq c_N(1+|\xi|)^{-N}$ for all N. Conversely, if $F(\zeta)$ is an entire function of exponential type A, and $F(\xi)$ is rapidly decreasing, then $F = \hat{f}$ for some such function f.*

One thing to keep in mind with these and other Paley-Wiener theorems is that since you are dealing with analytic functions, some estimates imply other estimates. Therefore it is possible to characterize the Fourier transforms in several different, but ultimately equivalent, ways.

The argument we gave for $\hat{f}(\zeta)$ to be entire of exponential type A is valid under the hypotheses of P.W. 1 because we have

$$\int_{-A}^{A} |f(x)|\, dx \;\leq\; \left(\int_{-A}^{A} |f(x)|^2\, dx\right)^{1/2} \left(\int_{-A}^{A} 1\, dx\right)^{1/2}$$

$$\leq\; (2A)^{1/2} \left(\int_{-A}^{A} |f(x)|^2\right)^{1/2} < \infty$$

by the Cauchy-Schwartz inequality. Of course $\int_{-\infty}^{\infty} |\hat{f}(\xi)|^2\, d\xi < \infty$ by the Plancherel formula, so the first part of P.W. 1 is proved. Similarly, we have proved the first part of P.W. 2, since $f \in \mathcal{D} \subseteq \mathcal{S}$ so $\hat{f} \in \mathcal{S}$ is rapidly decreasing. The converse statements are more difficult, and we will only give a hint of the proof. In both cases we know by the Fourier inversion formula that there exists $f(x)$, namely

$$f(x) = \frac{1}{2\pi} \int_{-\infty}^{\infty} F(\xi) e^{-ix\xi}\, d\xi$$

such that $\hat{f}(\xi) = F(\xi)$. The hard step is to show that f is supported in $[-A, A]$. In other words, if $|x| > A$ we need to show $f(x) = 0$. To do this we argue that it is permissible to make the change of variable $\xi \to \xi + i\eta$ in the Fourier inversion formula, for any fixed η. Of course this requires a contour integration argument. The result is

$$f(x) = \int_{-\infty}^{\infty} F(\xi + i\eta) e^{-ix\xi} e^{x\eta}\, d\xi$$

for any fixed η. Then we let η go to infinity; specifically, if $x > A$, we let $\eta \to -\infty$ and if $x < -A$ we let $\eta \to +\infty$, so that $e^{x\eta}$ goes to zero. The fact that F is of exponential type A means $F(\xi + i\eta)$ grows at worst like $e^{A|\eta|}$, but $e^{x\eta}$ goes to zero faster, so

$$\lim_{\eta \to \pm\infty} \int_{-\infty}^{\infty} F(\xi + i\eta) e^{-ix\xi} e^{x\eta}\, d\xi = 0$$

and hence $f(x) = 0$ as required. This is of course an oversimplification of the argument, because we also have the ξ integral to contend with.

In describing the Paley-Wiener theorem characterizing Fourier transforms of distributions of compact support, we need a growth estimate that is a little more complicated than exponential type, because these Fourier transforms may have polynomial growth in the real variable—for example, they may be polynomials.

Theorem 7.2.3 *P.W. 3: Let f be a distribution supported in $[-A, A]$. Then $\hat{f}$ is an entire function satisfying a growth estimate*

$$|\hat{f}(\xi)| \leq c(1 + |\zeta|)^N e^{A|\eta|}$$

for some c and N. Conversely, if F is an entire function satisfying such a growth estimate then it is the Fourier transform of a distribution with support in $[-A, A]$.

As an example, take $f = \delta'(x - A)$. Then we know $\hat{f}(\xi) = -i\xi e^{iA\xi}$, hence $\hat{f}(\zeta) = -i\zeta e^{iA\zeta}$ (since it is analytic) and this satisfies the growth estimate with $N = 1$.

The argument for P.W. 3 is a bit more complicated than before. Suppose we choose a cut-off function ψ that is identically one on a neighborhood of $[-A, A]$ and is supported in $[-A - \epsilon, A + \epsilon]$. Then $\hat{f}(\zeta) = \langle f, \psi(x)e^{ix\zeta} \rangle$ and we can apply the boundedness estimate discussed in section 6.5 to obtain

$$|\hat{f}(\zeta)| \leq c \sum_{k=0}^{N} \left\| \left(\frac{d}{dx} \right)^k \left(\psi(x)e^{ix\zeta} \right) \right\|_\infty .$$

Since differentiation of $e^{ix\zeta}$ produces powers of ζ up to N, and $|e^{ix\zeta}| = e^{-x\eta}$ is bounded by $e^{(A+\epsilon)|\eta|}$ on the support of ψ, we obtain the estimate

$$|\hat{f}(\zeta)| \leq c(1 + |\zeta|)^N e^{(A+\epsilon)|\eta|}$$

which is not quite what we wanted, because of the ϵ. (We cannot eliminate ϵ by letting it tend to zero because the constant c depends on ϵ and may blow up in the process.) The trick for eliminating the ϵ is to make ϵ and ψ depend on η, so that the product $\epsilon|\eta|$ remains constant. The argument for the converse involves convolving with an approximate identity and using P.W. 2.

All three theorems extend in a natural way to n dimensions. Perhaps the only new idea is that we must explain what is meant by an analytic function of several complex variables. It turns out that there are many equivalent definitions, perhaps the simplest being that the function is continuous and analytic in each of the n complex variables separately. If f is a function on $\mathbb{R}^n$ supported in $|x| \leq A$, then the Fourier transform extends to $\mathbb{C}^n$,

$$\hat{f}(\zeta) = \mathcal{F}_x(e^{-x\eta}f(x))(\xi),$$

as an entire analytic function, and again it has exponential type A, meaning

$$|\hat{f}(\zeta)| \le ce^{A|\eta|}$$

since $e^{-x\cdot\eta}$ has maximum value $e^{A|\eta|}$ on $|x| \le A$ (take $x = -A\eta/|\eta|$). Then P.W. 1, 2, 3 are true as stated with the ball $|x| \le A$ replacing the interval $[-A, A]$.

For a different kind of Paley-Wiener theorem, which does not involve compact support, let's return to one dimension and consider the factor $e^{-x\eta}$ in the formula for $\hat{f}(\zeta)$. Note that this either blows up or decays, depending on the sign of $x\eta$. In particular, if $\eta \ge 0$, then $e^{-x\eta}$ will be bounded for $x \ge 0$. So if f is supported on $[0, \infty)$, then $\hat{f}(\zeta)$ will be well defined, and analytic, on the upper half-space $\eta > 0$ (similarly for f supported on $(-\infty, 0\}$ and the lower half-space $\eta < 0$).

Theorem 7.2.4 *P.W. 4 Let f be supported on $[0, \infty)$ with $\int_0^\infty |f(x)|^2\, dx < \infty$. Then $\hat{f}(\zeta)$ is an analytic function on $\eta > 0$ and satisfies the growth estimate*

$$\sup_{\eta>0} \int_{-\infty}^\infty |\hat{f}(\xi + i\eta)|^2 d\xi < \infty.$$

Furthermore, the usual Fourier transform $\hat{f}(\xi)$ is the limit of $\hat{f}(\xi + i\eta)$ as $\eta \to 0^+$ in the following sense:

$$\lim_{\eta\to 0^+} \int_{-\infty}^\infty |\hat{f}(\xi + i\eta) - \hat{f}(\xi)|^2 \, d\xi = 0.$$

Conversely, suppose $F(\xi)$ is analytic in $\eta > 0$ and satisfies the growth estimate

$$\sup_{\eta>0} \int_{-\infty}^\infty |f(\xi + i\eta)|^2 d\xi < \infty.$$

Then $F = \hat{f}$ for such a function f.

As an application of the Paley-Wiener theorems (specifically P.W. 3), let's show that the wave equation in n dimensions has the same maximum speed of propagation that we observed in section 5.3 for $n = 1, 2, 3$. This amounts to showing that the distributions

$$\mathcal{F}^{-1}(\cos kt|\xi|) \qquad \text{and} \qquad \mathcal{F}^{-1}\left(\frac{\sin kt|\xi|}{k|\xi|}\right)$$

are supported in $|x| \le k|t|$. To apply P.W. 3 we need first to understand how the functions $\cos kt|\xi|$ and $\sin kt|\xi|/k|\xi|$ may be extended to entire analytic functions. Note that it will not do to replace $|\xi|$ by $|\zeta|$ since $|\zeta|$ is not analytic.

The key observation is that since both $\cos z$ and $\sin z/z$ are even functions of z, we can take $\cos\sqrt{z}$ and $\sin\sqrt{z}/\sqrt{z}$ and these will be entire analytic functions of z, even though $\sqrt{z}$ is not entire (it is not single-valued). Then the desired analytic functions are

$$\cos kt(\zeta\cdot\zeta)^{1/2}\qquad\text{and}\qquad\frac{\sin kt(\zeta\cdot\zeta)^{1/2}}{k(\zeta\cdot\zeta)^{1/2}}.$$

Indeed they are entire analytic, being the composition of analytic functions ($\zeta\cdot\zeta$ is clearly analytic), and they assume the correct values when ζ is real.

To apply P.W. 3 we need to estimate the size of these analytic functions; in particular we will show $|F(\zeta)| \le ce^{kt|\eta|}$ for each of them. The starting point is the observation

$$\begin{aligned}
|\cos(u+iv)| &= \frac{1}{2}|e^{i(u+iv)} + e^{-i(u+iv)}| \\
&\le \frac{1}{2}(e^{-v} + e^{v}) \le e^{|v|}
\end{aligned}$$

and similarly

$$\left|\frac{\sin(u+iv)}{u+iv}\right| \le e^{|v|}$$

(this requires a separate argument for $|u+iv| \le 1$ and $|u+iv| \ge 1$). To apply these estimates to our functions we need to write $(\zeta\cdot\zeta)^{1/2} = u+iv$ (this does not determine u and v uniquely, since we can multiply both by -1, but it *does* determine $|v|$ uniquely), and then we have $|F(\zeta)| \le ce^{|ktv|}$ for both functions. To complete the argument we have to show $|v| \le |\eta|$. This follows by algebra, but the argument is tricky. We have $\zeta\cdot\zeta = (u+iv)^2$, which means

$$\begin{aligned}
u^2 - v^2 &= |\xi|^2 - |\eta|^2 \\
uv &= \xi\cdot\eta
\end{aligned}$$

by equating real and imaginary parts, and of course $|\xi\cdot\eta| \le |\xi|\,|\eta|$ so we can replace the second equation by the inequality

$$u^2v^2 \le |\xi|^2|\eta|^2.$$

Using the first equation to eliminate u^2, we obtain the quadratic inequality

$$v^2(v^2 + |\xi|^2 - |\eta|^2) \le |\xi|^2|\eta|^2.$$

Completing the square transforms this into

$$\left(v^2 + \frac{|\xi|^2 - |\eta|^2}{2}\right)^2 \leq \left(\frac{|\xi|^2 + |\eta|^2}{2}\right)^2$$

and taking the square root yields $v^2 \leq |\eta|^2$ as desired.

So P.W. 3 says that $\mathcal{F}^{-1}(\cos kt|\xi|)$ and $\mathcal{F}^{-1}(\sin kt|\xi|/k|\xi|)$ are distributions supported in $|x| \leq kt$, confirming our calculations for $n = 1, 2, 3$. In fact, in all cases these distributions have been calculated explicitly, but the same argument works for more general hyperbolic differential equations.

7.3   The Poisson summation formula

We have computed the Fourier transform of δ to be the function that is identically 1. We can also compute the Fourier transform of any translate of δ,

$$\mathcal{F}_x\left(\delta(x - y)\right)(\xi) = e^{iy \cdot \xi}.$$

For simplicity let's assume $n = 1$. Let y vary over the integers (call it k) and sum:

$$\mathcal{F}\left(\sum_{k=-\infty}^{\infty} \delta(x - k)\right) = \sum_{k=-\infty}^{\infty} e^{ik\xi}.$$

The sum on the left (before taking the Fourier transform) defines a legitimate tempered distribution, which we can think of as an infinite "comb" of evenly spaced point masses. The sum on the right looks like a complicated function. In fact, it isn't. What it turns out to be is almost exactly the same as the sum on the left! This incredible fact is the Poisson summation formula.

Of course the sum $\sum_{k=-\infty}^{\infty} e^{ik\xi}$ does not exist in the usual sense, so we will not discover the Poisson summation formula by a direct attack. We will take a more circular route, starting with the idea of periodization. Given a function f on the line, we can create a periodic function (of period 2π) by the recipe

$$Pf(x) = \sum_{k=-\infty}^{\infty} f(x - 2\pi k)$$

assuming f has compact support or decays rapidly enough at infinity.

The key question to ask is the following: what is the relationship between the coefficients of the Fourier series expansion of Pf, $Pf(x) = \Sigma c_k e^{ikx}$ and the Fourier transform of f? Once you ask this question, the answer is not hard to find. We know

$$c_k = \frac{1}{2\pi} \int_0^{2\pi} Pf(x)e^{-ikx}\,dx.$$

If we substitute the definition of Pf, and interchange the sum and the integral (this is certainly valid if f has compact support), we find

$$
\begin{aligned}
c_k &= \frac{1}{2\pi} \sum_{j=-\infty}^{\infty} \int_0^{2\pi} f(x - 2\pi j)e^{-ikx}\,dx \\
&= \frac{1}{2\pi} \sum_{j=-\infty}^{\infty} \int_{-2\pi j}^{-2\pi(j-1)} f(x)e^{-ikx}\,dx \\
&= \frac{1}{2\pi} \int_{-\infty}^{\infty} f(x)e^{-ikx}\,dx \\
&= \frac{1}{2\pi} \hat{f}(-k)
\end{aligned}
$$

since $e^{-2\pi ikj} = 1$. So the Fourier series coefficients of Pf are essentially obtained by restricting $\hat{f}$ to the comb.

The Poisson summation formula now emerges if we compute $Pf(0)$ in two ways, from the definition, and by summing the Fourier series:

$$\sum_{k=-\infty}^{\infty} f(2\pi k) = \frac{1}{2\pi} \sum_{k=-\infty}^{\infty} \hat{f}(k).$$

The left side is just

$$\left\langle \sum_{k=-\infty}^{\infty} \delta(x - 2\pi k), f \right\rangle$$

and the right side is

$$\frac{1}{2\pi} \left\langle \sum_{k=-\infty}^{\infty} \delta(x - k), \hat{f} \right\rangle = \frac{1}{2\pi} \left\langle \mathcal{F}\left(\sum_{k=-\infty}^{\infty} \delta(x - k) \right), f \right\rangle$$

for $f \in \mathcal{S}$, so we have

$$\mathcal{F}\left(\sum_{k=-\infty}^{\infty} \delta(x - k) \right) = 2\pi \sum_{k=-\infty}^{\infty} \delta(x - 2\pi k).$$

In n dimensions the argument is similar. Let $\mathbb{Z}^n$ denote the lattice of points $(k_1, \ldots, k_n)$ with integer coordinates. Then we have

$$\mathcal{F}\left(\sum_{k \in \mathbb{Z}^n} \delta(x - k) \right) = (2\pi)^n \sum_{k \in \mathbb{Z}^n} \delta(x - 2\pi k)$$

and

$$\sum_{k \in \mathbb{Z}^n} f(2\pi k) = \frac{1}{(2\pi)^n} \sum_{k \in \mathbb{Z}^n} \hat{f}(k).$$

For which functions f does this second form hold? (Of course this is the form that Poisson discovered more than a century before distribution theory.) Certainly for $f \in \mathcal{S}$, but for many applications it is necessary to apply it to more general functions, and this usually requires a limiting argument. Unfortunately, it is not universally valid; there are examples of functions f for which both series converge absolutely, but to different numbers. A typical number theory application involves taking $\hat{f}$ to be the characteristic function of the ball $|x| \leq R$. Then the right side gives the number of lattice points inside the ball or, equivalently, the number of representations of all integers $\leq R^2$ as the sum of n squares. This number is, to first-order approximation, the volume of the ball. By using the Poisson summation formula, it is possible to find estimates for the difference.

The Poisson summation formula yields all sorts of fantastic results simply by substituting functions f whose Fourier transform can be computed explicitly. For example, take a Gaussian:

$$\sum_{k=-\infty}^{\infty} e^{-4\pi^2 t k^2} = \frac{1}{\sqrt{4\pi t}} \sum_{k=-\infty}^{\infty} e^{-k^2/4t}.$$

This is an important identity in the theory of θ-functions.

I would like to consider now a crystalographic interpretation of the Poisson summation formula. We can regard $\sum_{k \in \mathbb{Z}^n} \delta(x - k)$ as a model of a cubic crystal, each term $\delta(x - k)$ representing an atom located at the lattice point k. The Fourier transform is then a model of the output of an X-ray diffraction experiment. Such experiments do in fact produce the characteristic pattern of bright spots on a cubic lattice, even though the real crystal is of finite spatial extent. Since not all crystals are cubic, this point of view leads us to seek a slight generalization of the Poisson summation formula: we want to compute the Fourier transform of $\sum_{\gamma \in \Gamma} \delta(x - \gamma)$ where Γ is an arbitrary lattice in $\mathbb{R}^n$. By definition, a lattice is a set represented by $\gamma = k_1 v_1 + \cdots + k_1 v_n$ where k_j are integers and $v_1, \ldots, v_n$, are linearly independent vectors in $\mathbb{R}^n$. From this description we see that $\Gamma = A\mathbb{Z}^n$ where A is an invertible $n \times n$ matrix. (A small point that may cause confusion is that the vectors $v_1, \ldots, v_n$ are not unique, nor is the matrix A. For example, the lattice in $\mathbb{R}^2$ generated by $v_1 = (0, 1)$ and $v_2 = (1, 1)$ is the same as $\mathbb{Z}^2$.)

Given any lattice Γ, the *dual lattice* Γ' is defined to be the set of vectors $x \in \mathbb{R}^n$ such that $x \cdot \gamma = 2\pi \cdot$ integer for every $\gamma \in \Gamma$. For example, if $\Gamma = \mathbb{Z}^n$, then $\Gamma' = 2\pi\mathbb{Z}^n$. More generally, if $\Gamma = A\mathbb{Z}^n$ then $\Gamma' = 2\pi(A^{tr})^{-1}\mathbb{Z}^n$ since $(A^{tr})^{-1}x \cdot Ay = x \cdot y$. We claim that

$$\mathcal{F}\left(\sum_{\gamma \in \Gamma} \delta(x - \gamma)\right) = c_\Gamma \sum_{\gamma' \in \Gamma'} \delta(x - \gamma')$$

is the Poisson summation formula for general lattices, where the constant c_Γ can be identified with the volume of the fundamental domain of Γ'. The idea is that we have

$$\begin{aligned}
\mathcal{F}\left(\sum_{\gamma \in \Gamma} \delta(x - \gamma)\right)(\xi) &= \sum_{\gamma \in \Gamma} e^{i\xi \cdot \gamma} \\
&= \sum_{k \in \mathbb{Z}^n} e^{i\xi \cdot Ak} = \sum_{k \in \mathbb{Z}^n} e^{i(A^{tr}\xi) \cdot k} \\
&= (2\pi)^n \sum_{k \in \mathbb{Z}^n} \delta(A^{tr}\xi - k)
\end{aligned}$$

by the usual Poisson summation formula. But $\delta(A^{tr}\xi - k) = |\det A|^{-1}\delta(\xi - (A^{tr})^{-1}k)$, which proves our formula with $c_\Gamma = (2\pi)^n|\det A|^{-1}$.

We can now give a brief introduction to the notion of *quasicrystals*. The idea is that there exist nonperiodic arrangements of atoms that produce discrete Fourier transforms. The particular arrangement we describe is obtained by strip-projection from a higher dimensional lattice. To be specific, take a lattice Γ in $\mathbb{R}^2$ obtained from $\mathbb{Z}^2$ by rotation through an "irrational" angle (more precisely, it is the tangent of the angle that must be an irrational number). We take a strip $|y| \le b$ parallel to the x-axis and then project the lattice points in this strip onto the x-axis; shown in the figure. The result is just $\sum_{|\gamma_2| \le b} \delta(x - \gamma_1)$ where $\gamma = (\gamma_1, \gamma_2)$ varies over Γ. This is clearly a sum of discrete "atoms," and it will not be periodic. The Fourier transform is $\sum_{|\gamma_2| \le b} e^{i\gamma_1 \xi}$ and this can be given a form similar to the Poisson summation formula. We start with $\sum_\Gamma e^{i\gamma_1 \xi} e^{i\gamma_2 \eta} = (2\pi)^2 \sum_{\Gamma'} \delta(\xi - \gamma_1', \eta - \gamma_2')$, multiply both sides by a function $g(\eta)$, and integrate to obtain

$$\sum_\Gamma e^{i\gamma_1 \xi} \hat{g}(\gamma_2) = (2\pi)^2 \sum_{\Gamma'} g(\gamma_2')\delta(\xi - \gamma_1').$$

Clearly we want to choose the function g such that $\hat{g}(t) = \chi(|t| \le b)$. Since we know $\int_{-b}^b e^{its}dt = 2\sin sb/s$ it follows from the Fourier inversion formula that $g(s) = (\sin sb)/\pi s$ is the correct choice. This yields

$$\sum_{|\gamma_2| \le b} e^{i\gamma_1 \xi} = 4\pi \sum_{\Gamma'} \frac{\sin b\gamma_2'}{\gamma_2'}\delta(\xi - \gamma_1')$$

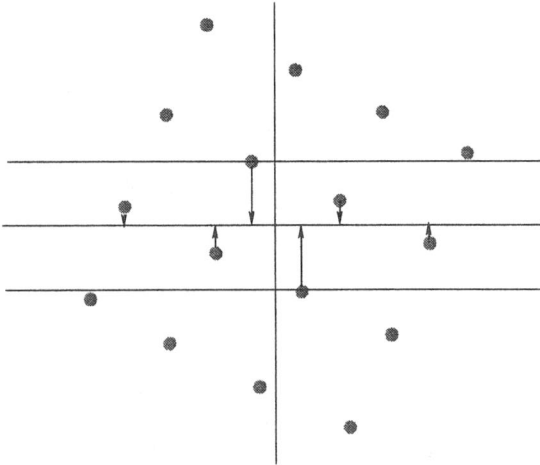

Figure 7.1

as the Fourier transform. This is a weighted sum of δ-functions centered at the points γ_1' which are the first coordinates of the dual lattice.

Despite the superficial resemblance, there are some differences between this quasi-periodic Poisson summation formula and the usual one. The main difference is that the points γ_1' are not isolated; in fact they form a dense subset of the line. Of course the contribution from lattice points with large γ_2' will be multiplied by the small coefficient $(\sin b\gamma_2')/\gamma_2'$, so we can imagine the Fourier transform as a discrete set of relatively bright stars amid a Milky Way of stars too dim to perceive individually. The trouble is that this image is not quite accurate, because of the slow decay of $g(s)$. The sum defining the Fourier transform is not absolutely convergent, so our Milky Way is not infinitely bright only because of cancellation between positive and negative stars. This difficulty can be overcome by choosing a smoother cut-off function as you approach the boundary of the strip. This leads to a different choice of g, one with faster decay.

7.4 Probability measures and positive definite functions

A probability measure μ on $\mathbb{R}^n$ is an assignment of a probability to all reasonable subsets of $\mathbb{R}^n$ which is positive and additive: $0 \leq \mu(A) \leq 1$ and $\mu(A \cup B) = \mu(A) + \mu(B)$ if A and B are disjoint. In addition we require the conditions $\mu(\phi) = 0, \mu(\mathbb{R}^n) = 1$, where ϕ denotes the empty set, and a technical condition called countable additivity:

$$\mu \left(\bigcup_{j=1}^{\infty} A_j \right) = \sum_{j=1}^{\infty} \mu(A_j)$$

if the sets A_j are disjoint. All the intuitive examples of probability distributions on $\mathbb{R}^n$ satisfy these conditions, although it is not always easy to verify them. Associated to a probability measure μ is an integral $\int f d\mu$ or $\int f(x) d\mu(x)$ defined for continuous functions (and more general functions as well) which gives the "expectation" of the function with respect to the probability measure. For continuous functions with compact support, the integral may be formed in the usual way, as the limit of sums $\sum f(x_j)\mu(I_j)$ where $x_j \in I_j$ and $\{I_j\}$ is a partition of $\mathbb{R}^n$ into small boxes. Another way of thinking about it is to use a "Monte Carlo" approximation: choose points $y_1, y_2, \ldots$ at random according to the probability measure and take $\lim_{N \to \infty} \frac{1}{N} (f(y_1) + \cdots + f(y_N))$.

A probability measure can be regarded as a distribution, namely $\langle \mu, \varphi \rangle = \int \varphi d\mu$, and as we have seen in section 6.4, it is a positive distribution, meaning $\langle \mu, \varphi \rangle \geq 0$ if $\varphi \geq 0$. The condition $\mu(\mathbb{R}^n) = 1$ is equivalent to $\int 1 d\mu = 1$, which we can write $\langle \mu, 1 \rangle = 1$ by abuse of notation (the constant function 1 is not in $\mathcal{D}$, so $\langle \mu, 1 \rangle$ is not defined a priori). Of course this means $\lim_{k \to \infty} \langle \mu, \psi_k \rangle = 1$ where ψ_k is a suitable sequence of test functions approximating the constant function 1. The converse statement is also true: any positive distribution with $\langle \mu, 1 \rangle = 1$ is associated to a probability measure in this way. If f is a positive, integrable function with $\int f(x) dx = 1$ (ordinary integration), then there is an associated probability measure, denoted $f(x) dx$, defined by $\mu(A) = \int_A f(x) dx$, whose associated integral is just $\int \varphi d\mu = \int \varphi(x)f(x) dx$. Such measures are called *absolutely continuous*. Other examples are discrete measures $\mu = \sum p_j \delta(x - a_j)$ where $\{p_j\}$ are discrete probabilities

$$\left(0 \leq p_j \leq 1 \quad \text{and} \quad \sum p_j = 1 \right)$$

distributed on the points a_j (the sum may be finite or countable) and more exotic Cantor measures described in section 6.4.

Probability measures have Fourier transforms (the associated distributions are tempered), which are given as functions directly by the formula

$$\hat{\mu}(\xi) = \langle \mu, e^{ix \cdot \xi} \rangle = \int e^{ix \cdot \xi} \, d\mu(x).$$

These functions are continuous (even uniformly continuous) and bounded; in fact,

$$|\hat{\mu}(x)| \leq \int 1 \, d\mu(x) = 1$$

and

$$\hat{\mu}(0) = \int 1 \, d\mu(x) = 1.$$

If $\mu = f \, dx$ then $\hat{\mu} = \hat{f}$ in the usual sense. In general, $\hat{\mu}$ does not vanish at infinity, as for example $\hat{\delta} = 1$.

It turns out that we can exactly characterize the Fourier transforms of probability measures. This characterization is called *Bochner's theorem* and involves the notion of *positive definiteness*. Before describing it, let me review the analogous concept for matrices.

Let A_{jk} denote an $N \times N$ matrix with complex entries. Associated to this matrix is a quadratic form on $\mathbb{C}^N$, which we define by $\langle Au, u \rangle = \sum_{j,k} A_{jk} \bar{u}_j u_k$ for $u = (u_1, \ldots, u_N)$ a vector in $\mathbb{C}^N$. We say the matrix is *positive definite* if the quadratic form is always nonnegative, $\langle Au, u \rangle \geq 0$ for all u. Note that, a priori, the quadratic form is not necessarily real, but this is required when we write $\langle Au, u \rangle \geq 0$. Also note that the term *nonnegative definite* is sometimes used, with "positive definite" being reserved for the stronger condition

$$\langle Au, u \rangle > 0 \quad \text{if} \quad u \neq 0.$$

For our purposes it is better not to insist on this strict positivity. Also, it is sometimes assumed that the matrix is Hermitian, meaning $A_{jk} = \bar{A}_{kj}$. In the situation we are considering this will always be the case, although it is not necessary to insist on it. Now we can define what is meant by a *positive definite function* on $\mathbb{R}^n$.

Definition 7.4.1 *A continuous (complex-valued) function F on $\mathbb{R}^n$ is said to be positive definite if the matrix $A_{jk} = F(x_j - x_k)$ is positive definite for any choice of points $x_1, \ldots, x_N$ in $\mathbb{R}^n$, and any N. Specifically, this means*

$$\sum_j \sum_k F(x_j - x_k) \bar{u}_j u_k \geq 0$$

for any choice of $u \in \mathbb{C}^N$.

Now we claim that the Fourier transform of a probability measure is a positive definite function. This is very easy to see, since

$$\sum_j \sum_k \hat{\mu}(\xi_j - \xi_k)\bar{u}_j u_k \;=\; \sum_j \sum_k \left(\int e^{ix\cdot(\xi_j - \xi_k)} d\mu(x) \right) \bar{u}_j u_k$$

$$=\; \int \left(\sum_j \sum_k e^{ix\cdot\xi_j}\bar{u}_j e^{-ix\cdot\xi_k} u_k \right) d\mu(x)$$

$$=\; \int \left| \sum_k e^{-ix\cdot\xi_k} u_k \right|^2 d\mu(x)$$

and the integral of a nonnegative function is nonnegative (note that it might be zero if the probability measure is concentrated on the points where the function $\sum_k e^{-ix\cdot\xi_k} u_k$ vanishes). Thus we have established the easy half of

Theorem 7.4.2 *Bochner's Theorem: A function F is the Fourier transform of a probability measure, if and only if*

1. *F is continuous.*

2. *$F(0) = 1$.*

3. *F is positive definite.*

To prove the converse, we first need to transform the positive definite condition from a discrete to a continuous form. The idea is that, in the condition defining positive definiteness, we can think of the values u_j as weights associated to the points x_j. Then the sum approximates a double integral. The continuous form should thus be

$$\iint F(x - y)\overline{\varphi(x)}\varphi(y)\, dx\, dy \geq 0$$

for any test function $\varphi \in \mathcal{D}$. A technical lemma, whose proof we will omit, asserts that this condition is in fact equivalent to positive definiteness (for continuous functions).

So let us start with a continuous, positive definite function F. The positive definite condition implies that F is bounded, so that $F = \widehat{T}$ for some tempered distribution T. We are going to prove that T is in fact a positive distribution. In fact, the continuous form of positive definiteness says exactly $\langle F * \varphi, \bar{\varphi} \rangle \geq 0$ for every $\varphi \in \mathcal{D}$, and the same holds for every $\varphi \in \mathcal{S}$ because $\mathcal{D}$ is dense in $\mathcal{S}$. Now a change of variable lets us write this as

$$\langle F, \tilde{\bar{\varphi}} * \varphi \rangle \geq 0$$

where $\tilde{\varphi}(x) = \overline{\varphi(-x)}$. Since $F = \hat{T}$ we have

$$\langle F, \tilde{\varphi} * \varphi \rangle = \langle \hat{T}, \tilde{\varphi} * \varphi \rangle = \langle T, \mathcal{F}(\tilde{\varphi} * \varphi) \rangle = \langle T, |\hat{\varphi}|^2 \rangle.$$

So the positive definiteness of F says that T is nonnegative on functions of the form $|\hat{\varphi}|^2$ for all $\varphi \in \mathcal{S}$, which is the same a $|\varphi|^2$ for all $\varphi \in \mathcal{S}$, since the Fourier transform is an isomorphism of $\mathcal{S}$.

We have almost proved what we claimed, because to say T is a positive distribution is to say $\langle T, \psi \rangle \geq 0$ for any test function ψ that is nonnegative. It is merely a technical matter to pass from functions of the form $|\varphi|^2$ to general nonnegative functions. Once we know that T is a positive distribution, by the results of section 6.4 it comes from a positive measure μ (not necessarily finite, however). From $\hat{T}(0) = 1$ and $\hat{T}$ continuous we conclude that $\mu(\mathbb{R}^n) = \hat{T}(0) = 1$, so μ is a probability measure.

Positive definiteness is not always an easy condition to verify, but here is one example where it is possible: take F to be a Gaussian. Since we know F is the Fourier transform of a Gaussian, which is a positive function, we know from Bochner's theorem that $e^{-t|x|^2}$ is positive definite. We can verify this directly, first in one dimension. Then

$$\sum_j \sum_k F(x_j - x_k)\bar{u}_j u_k = \sum_j \sum_k e^{-t|x_j - x_k|^2} \bar{u}_j u_k$$

$$= \sum_j \sum_k e^{-tx_j^2} e^{-tx_k^2} e^{2tx_j x_u} \bar{u}_j u_k.$$

The term $e^{2tx_j x_k}$ seems to present difficulties, so let's expand it in a power series,

$$e^{2tx_j x_k} = \sum_{m=0}^{\infty} \frac{(2t)^m}{m!} x_j^m x_k^m.$$

When we substitute this into our computation we can take the m-summation out front to obtain

$$\sum_{m=0}^{\infty} \frac{(2t)^m}{m!} \sum_j \sum_k e^{-tx_j^2} e^{-tx_k^2} x_j^m x_k^m \bar{u}_j u_k$$

$$= \sum_{m=0}^{\infty} \frac{(2t)^m}{m!} \left| \sum_k e^{-tx_k^2} x_k^m u_k \right|^2$$

which is clearly positive. The argument in n dimensions is similar, but we have to first break up $e^{2tx_j \cdot x_k}$ into a product of n component functions (the notation is awkward here, since x_j and x_k are vectors in $\mathbb{R}^n$) and then take the power series expansion of each exponential factor.

7.5   The Heisenberg uncertainty principle

One of the most famous and paradoxical predictions of quantum theory is the statement that it is impossible to determine simultaneously both the position and momentum of a particle with extreme accuracy. The quantitative version of this statement is the Heisenberg uncertainty principle, which may be formulated as an inequality in Fourier analysis (it can also be expressed in terms of operator theory, as we will discuss later). To describe this inequality, let us recall briefly the basic principles of quantum mechanics. A particle is described by a wave function $\psi(x)$, which is a complex-valued function on $\mathbb{R}^n$ satisfying $\int |\psi(x)|^2 \, dx = 1$ (strictly speaking, the wave function is only determined up to a phase, i.e., $\psi(x)$ and $e^{i\alpha}\psi(x)$ represent the same particle). The probability measure $|\psi(x)|^2 \, dx$ gives the probability of finding the particle in some region of space: prob (particle lies in A) $= \int_A |\psi(x)|^2 \, dx$. The mean position of the particle is then $\bar{x}$, given by $\bar{x}_j = \int x_j |\psi(x)|^2 \, dx$, and the variance of the probability measure,

$$\mathrm{Var} = \int |x - \bar{x}|^2 |\psi(x)|^2 \, dx$$

is a good index of the amount of "spread" of the measure, hence the "uncertainty" in measuring the position of the particle. We will write $\mathrm{Var}(\psi)$ to indicate the dependence on the wave function. An important but elementary observation is that the variance can be expressed directly as

$$\mathrm{Var}(\psi) = \inf_{y \in \mathbb{R}^n} \int |x - y|^2 |\psi(x)|^2 \, dx$$

without first defining the mean. To see this, write $y = \bar{x} + z$. Then

$$\int |x - \bar{x} - z|^2 |\psi(x)|^2 dx = \int |x - \bar{x}|^2 |\psi(x)|^2 + |z|^2 \int |\psi(x)|^2 dx$$
$$-2 \int (x - \bar{x}) \cdot z |\psi(x)|^2 dx.$$

We claim the last integral vanishes, because $(x-\bar{x}) \cdot z = \sum_{j=1}^n (x_j - \bar{x}_j) z_j$ and $\int (x_j - \bar{x}_j)|\psi(x)|^2 \, dx = 0$ by the definition of $\bar{x}_j$ (remember $\int |\psi(x)|^2 \, dx = 1$). Thus

$$\int |x - \bar{x} - z|^2 |\psi(x)|^2 \, dx = \mathrm{Var}(\psi) + |z|^2$$

and the minimum is clearly assumed when $z = 0$.

In quantum mechanics, observables are represented by operators (strictly speaking, Hermitian operators). If A is such an operator, then $\langle A\psi, \psi \rangle =$

$\int A\psi(x)\overline{\psi(x)}\,dx$[1] represents the expected value of the corresponding observable on the particle represented by ψ. In other words, this is the average value you would actually measure for the observable on a collection of identical particles described by ψ. If we write $\bar{A} = \langle A\psi, \psi \rangle$, then

$$\text{Var}(A) = \langle (A - \bar{A})\psi, (A - \bar{A})\psi \rangle = \int |A\psi(x) - \bar{A}\psi(x)|^2 \, dx$$

represents the variance, or uncertainty, in the measurement of A. For the observable "position," the operator A is multiplication by x (actually this is a vector of operators, since position is a vector observable). You can easily verify that $\bar{A}$ and $\text{Var}(A)$ correspond to $\bar{x}$ and $V(\psi)$, which we defined before.

For the observable "momentum," the operator is a multiple of

$$i\frac{\partial}{\partial x} = \left(i\frac{\partial}{\partial x_1}, \ldots, i\frac{\partial}{\partial x_n} \right).$$

We will not be concerned here with the constants in the theory, including the famous Planck's constant. For this operator, the computation of mean and variance can be moved to the Fourier transform side (this was the way we described it in section 5.4). For the mean, we have

$$\int i\frac{\partial \psi}{\partial x_j}(x)\overline{\psi(x)}\,dx = \frac{1}{(2\pi)^n}\int \left(i\frac{\partial \psi}{\partial x_j} \right)\hat{}(\xi)\overline{\hat{\psi}(\xi)}\,d\xi$$

$$= \frac{1}{(2\pi)^n}\int \xi_j|\hat{\psi}(\xi)|^2\,d\xi$$

by the Plancherel formula and the fact that $(\partial\psi/\partial x_j)\hat{}(\xi) = -i\xi_j\hat{\psi}(\xi)$. Note that the constant $(2\pi)^{-n}$ is easily interpreted as follows: $(2\pi)^n = \int |\hat{\psi}(\xi)|^2\,d\xi$, so the function $(2\pi)^{-n/2}\hat{\psi}(\xi)$ has the correct normalization

$$\int |(2\pi)^{-n/2}\hat{\psi}(\xi)|^2\,d\xi = 1,$$

and so the mean of the momentum observable for ψ is the same as the mean of the position variable for $(2\pi)^{-n/2}\hat{\psi}$. Similarly, the variance for momentum is $\text{Var}((2\pi)^{-n/2}\hat{\psi})$.

By choosing ψ very concentrated around $\bar{x}$, we can make $\text{Var}(\psi)$ as small as we please. In other words, quantum mechanics does not preclude the existence of particles whose position is very localized. Similarly, there are

---

[1] Note that in this section we are using the complex conjugate in the inner product, contrary to previous usage.

also particles whose momentum is very localized ($\hat{\psi}$ is concentrated). What the Heisenberg uncertainty principle asserts is that we cannot achieve both localizations simultaneously for the same wave function. The reason is that the product of the variances is bounded below by $n^2/4$; in other words,

$$\text{Var}(\psi)\text{Var}((2\pi)^{-n/2}\hat{\psi}) \geq n^2/4.$$

Thus if we try to make $\text{Var}(\psi)$ very small, we are forced to make $\text{Var}((2\pi)^{-n/2}$ $\hat{\psi})$ large, and vice versa.

To prove the Heisenberg uncertainty principle, consider first the case $n = 1$. Let us write A for the operator $i(d/dx)$ and B for the operator of multiplication by x. Note that these operators do not commute. If we write $[A, B] = AB - BA$ for the commutator, then we find

$$[A, B]\psi = i\frac{d}{dx}(x\psi) - xi\frac{d}{dx}\psi = i\psi,$$

or simply $-i[A, B] = I$ (the identity operator). Also, both operators are Hermitian, $\langle A\psi_1, \psi_2\rangle = \langle \psi_1, A\psi_2\rangle$ by integration by parts (the i factor produces a minus sign under complex conjugation, canceling the minus sign from integration by parts), and similarly

$$\langle B\psi_1, \psi_2\rangle = \langle \psi_1, B\psi_2\rangle.$$

The commutation identity thus yields

$$
\begin{aligned}
1 &= -i\langle(AB - BA)\psi, \psi\rangle = -i\langle B\psi, A\psi\rangle + i\langle A\psi, B\psi\rangle \\
&= 2\,\text{Re}(i\langle A\psi, B\psi\rangle).
\end{aligned}
$$

Then the Cauchy-Schwartz inequality gives

$$1 \leq 2|\langle A\psi, B\psi\rangle| \leq 2\langle A\psi, A\psi\rangle^{1/2}\langle B\psi, B\psi\rangle^{1/2}$$

or

$$\langle A\psi, A\psi\rangle\langle B\psi, B\psi\rangle \geq \frac{1}{4}$$

which is exactly the Heisenberg uncertainty principle in the special case where the means $\bar{A}$ and $\bar{B}$ are both zero, for then $\langle A\psi, A\psi\rangle = \int |A\psi|^2\, dx$ is the variance of momentum, and $\langle B\psi, B\psi\rangle = \int |B\psi|^2\, dx$ is the variance of position.

However, the general case is easily reduced to this special case. One way to see this is to observe that translating ψ moves the mean position without changing the momentum calculation ($\hat{\psi}$ is multiplied by an exponential of absolute value one, so $|\hat{\psi}|^2$ is unchanged). Similarly, we may translate $\hat{\psi}$

(multiply ψ by e^{-iax}) without changing the position computation. Thus applying the above argument to $e^{-iax}\psi(x-b)$ for $a = \bar{A}$ and $b = \bar{B}$ yields a wave function with both means zero, and the above argument applied to this wave function yields the desired Heisenberg uncertainty principle for ψ. Another way to see the same thing is to observe that the operators $A - \bar{A}I$ and $B - \bar{B}I$ satisfy the same commutation relation, so the above argument applied to these operators yields

$$\langle A\psi - \bar{A}\psi, A\psi - \bar{A}\psi \rangle \langle B\psi - \bar{B}\psi, B\psi - \bar{B}\psi \rangle \geq \frac{1}{4}$$

which is the general form of the uncertainty principle.

Now there are two important lessons to be drawn from our argument. The first is that this is really an operator theoretic result. We only used the facts that A and B were Hermitian operators whose commutator satisfied the identity $-i[A,B] = I$. In fact, the same argument works if we have the inequality $-i[A,B] \geq I$, which is sometimes useful in applications (this just means $\langle -i[A,B]\psi, \psi \rangle \geq \langle \psi, \psi \rangle$). Two such operators are called *complementary pairs* in quantum mechanics. Thus, the operator version of the uncertainty principle asserts $\mathrm{Var}(A) \cdot \mathrm{Var}(B) \geq \frac{1}{4}$ for complementary pairs. In fact, to get the Fourier analytic form of the uncertainty principle we needed to use the identity

$$\mathrm{Var}(A) = \mathrm{Var}((2\pi)^{-1/2}\hat{\psi})$$

for the momentum operator $A = i(d/dx)$.

The second important lesson that emerges from the proof is that we can also determine exactly when the inequality in the uncertainty principle is an equality. In the special case that both operators have mean zero, the only place we used an inequality was in the estimate

$$2\,\mathrm{Re}(i\langle A\psi, B\psi \rangle) \leq 2\langle A\psi, A\psi \rangle^{1/2}\langle B\psi, B\psi \rangle$$

which involved the Cauchy-Schwartz inequality. But we know that this is an equality if and only if the two functions $A\psi$ and $B\psi$ are proportional to each other (in order to have equality above we also need to know that $i\langle A\psi, B\psi \rangle$ is real and positive). But this leads to a first-order differential equation $\psi' = cx\psi$ which has exactly the Gaussians as solutions, and it is easy to check that the condition $i\langle A\psi, B\psi \rangle \geq 0$ is satisfied for any Gaussian ce^{-tx^2}.

Thus we conclude that there is equality in the uncertainty principle exactly in the case that ψ is a Gaussian of arbitrary variance, translated arbitrarily in both space and momentum, i.e., $(2t/\pi)^{1/4}e^{iax}e^{-t|x-b|^2}$ (the constant is chosen to make $\int |\psi|^2\,dx = 1$). One could also compute the variances exactly in this case. An interesting point about these functions is that by varying the parameter t we may obtain all possible values for $\text{Var}(A)$ and $\text{Var}(B)$ whose product equals $\frac{1}{4}$.

For the Heisenberg uncertainty principle in n dimensions, we consider the n complementary pairs $A_j = i(\partial/\partial x_j)$ and $B_j = $ multiplication by x_j (it is easy to see $-i[A_j, B_j] = I$), and hence we have

$$\text{Var}(A_j)\text{Var}(B_j) \geq \frac{1}{4}$$

and it follows that

$$\left(\sum_j \text{Var}(A_j)\right) \left(\sum_J \text{Var}(B_j)\right) \geq n^2/4$$

by elementary inequalities; this is the n dimensional inequality. Again, equality holds exactly when ψ is a Gaussian, translated in space and momentum.

Although the proof of the uncertainty principle we gave was not Fourier analytic, the interpretation in Fourier analytic terms is extremely interesting. For one thing, it gives an interpretation of the uncertainty principle in classical physics. Suppose $\psi(t)$ is a sound wave. Then $\text{Var}(\psi)$ is an index of its concentration in time. We interpret $\hat{\psi}(t)$ as the frequency distribution of the sound wave, so $\text{Var}((2\pi)^{-1/2}\hat{\psi})$ is an index of the concentration of the pitch. The uncertainty principle then says that a sound cannot be very concentrated in both time and pitch. In particular, short duration tones will have a poorly determined pitch. String players and singers take advantage of this: in very rapid passages, errors in intonation will not be noticeable.

The Fourier analytic version of the uncertainty principle also has an important interpretation in signal processing. The idea is that one would love to have a basic signal that is very localized in both time and frequency. The uncertainty principle forbids this, and moreover tells you that the Gaussian is the best compromise. This leads to the *Gabor transform*. Take a fixed Gaussian, say $\psi(t) = \pi^{-1/4}e^{-t^2/2}$, translate it in time and frequency and take the inner product with a general signal $f(t)$ to be analyzed:

$$G(a,b) = \pi^{-1/4} \int_{-\infty}^{\infty} f(t)e^{-iat}e^{-(t-b)^2/2}dt.$$

This gives a "snapshot" of the strength of the signal in the vicinity of the time $t = b$ and in the vicinity of the frequency a. The rough localization achieved by the Gaussian is the best we can do in this regard, according to the uncertainty principle.

We can always recover the signal f from its Gabor transform, via the inversion formula

$$f(t) = \frac{1}{2\pi}\pi^{-1/4} \int_{-\infty}^{\infty} \int_{-\infty}^{\infty} G(a,b)e^{iat}e^{-(t-b)^2/2}\, da\, db.$$

To establish this formula, say for $f \in S$ (it holds more generally, in a suitable sense) we just substitute into the right side the definition of $G(a,b)$ (renaming the dummy variable s):

$$\frac{1}{2\pi}\pi^{-1/2} \int_{-\infty}^{\infty} \int_{-\infty}^{\infty} \int_{-\infty}^{\infty} f(s)e^{ia(t-s)}e^{-(t-b)^2/2}e^{-(s-b)^2/2}\, ds\, da\, db$$

and using the Fourier inversion formula for the s and a integrals we obtain

$$\pi^{-1/2} \int_{-\infty}^{\infty} f(t)e^{-(t-b)^2}\, db.$$

Since $\int_{-\infty}^{\infty} e^{-(t-b)^2}\, db = \pi^{1/2}$ we have $f(t)$ as claimed.

The inversion formula is perhaps not so surprising because the Gabor transform is highly redundant. We start with a signal that is a function of one variable and create a transform that is a function of two variables. One of the interesting questions in signal processing is whether we can recover the signal by "sampling" the Gabor transform; i.e., restricting the variables a and b to certain discrete choices.

7.6 Hermite functions

Gaussian functions have played a central role in our description of Fourier analysis, in part because the Fourier transform of a Gaussian is again a Gaussian. Specifically, if

$$f(x) = e^{-x^2/2} \quad \text{on} \quad \mathbb{R}^1$$

then

$$\mathcal{F}f = \sqrt{2\pi}f.$$

From a certain point of view, the Gaussian is just the first of a sequence of functions, called *Hermite functions*, all of them satisfying an analogous condition of being an eigenfunction of the Fourier transform (recall that an *eigenfunction* for any operator A is defined to be a nontrivial solution of $Af = \lambda f$ for a constant λ called the *eigenvalue*, where nontrivial means f is not identically zero). Because the Fourier inversion formula can be written

$\mathcal{F}^2\varphi(x) = 2\pi\varphi(-x)$ it follows that $\mathcal{F}^4\varphi = (2\pi)^2\varphi$, so the only allowable eigenvalues must satisfy $\lambda^4 = (2\pi)^2$, hence

$$\lambda = \sqrt{2\pi}, \sqrt{2\pi}i, -\sqrt{2\pi} \text{ or } -i\sqrt{2\pi}.$$

From a certain point of view, the problem of finding eigenfunctions for the Fourier transform has a trivial solution. If we start with any function φ, then the function

$$f = \varphi + (2\pi)^{-1/2}\mathcal{F}\varphi + (2\pi)^{-1}\mathcal{F}^2\varphi + (2\pi)^{-3/2}\mathcal{F}^3\varphi$$

satisfies $\mathcal{F}f = (2\pi)^{1/2}f$, and analogous linear combinations give the other eigenvalues. You might worry that f is identically zero, but we can recover φ as a linear combination of the 4 eigenfunctions, so at least one must be nontrivial. More to the point, such an expression gives us very little information about the eigenfunctions, so it is not considered very interesting.

It is best to present the Hermite functions not as eigenfunctions of the Fourier transform, but rather as eigenfunctions of the operator $H = -(d^2/dx^2) + x^2$. This operator is known as the *harmonic oscillator*, and the question of its eigenfunctions and eigenvalues (spectral theory) is important in the quantum mechanical theory of this system. Here we will work in one-dimension, since the n-dimensional theory of Hermite functions involves nothing more than taking products of Hermite functions of each of the coordinate variables.

The spectral theory of the harmonic oscillator H is best understood in terms of what the physicists call *creation* and *annihilation* operators $A^* = (d/dx) - x$ and $A = -(d/dx) - x$. It is easy to see that the creation operator A^* is the adjoint operator to the annihilation operator A. Now we need to do some algebra involving A^*, A, and H. First we need

$$A^*A = H - 1 \text{ and } AA^* = H + 1,$$

as you can easily check. Then $AH = AA^*A + A$ while $HA = AA^*A - A$ so $[A, H] = 2A$ and similarly $A^*H = A^*AA^* - A^*$ while $HA^* = A^*AA^* + A^*$ so $[A^*, H] = -2A^*$.

So what are the possible eigenvalues λ for H? Suppose $H\varphi = \lambda\varphi$. Then λ is real because H is self-adjoint:

$$\lambda\langle\varphi,\varphi\rangle = \langle H\varphi,\varphi\rangle = \langle\varphi,H\varphi\rangle = \bar{\lambda}\langle\varphi,\varphi\rangle,$$

hence $\lambda = \bar{\lambda}$. Furthermore, we claim $\lambda \geq 1$. To see this use $A^*A = H - 1$ to get

$$\lambda\langle\varphi,\varphi\rangle = \langle H\varphi,\varphi\rangle = \langle A^*A\varphi,\varphi\rangle + \langle\varphi,\varphi\rangle = \langle A\varphi,A\varphi\rangle + \langle\varphi,\varphi\rangle \geq \langle\varphi,\varphi\rangle,$$

and furthermore $\lambda = 1$ is possible only if $A\varphi = 0$. But $A\varphi = 0$ is a first-order differential equation whose unique solution (up to a constant multiple) is the Gaussian $e^{-x^2/2}$. Since $Ae^{-x^2/2} = 0$ we have

$$He^{-x^2/2} = e^{-x^2/2} + A^*Ae^{-x^2/2} = e^{-x^2/2}.$$

So we have identified 1 as the lowest eigenvalue (the *bottom of the spectrum*) with multiplicity one, and the Gaussian as the unique corresponding eigenfunction (the *groundstate*). To get a hold of the rest of the spectrum we need a little lemma that describes the action of the creation and annihilation operators on eigenfunctions:

Lemma 7.6.1 *Suppose φ is an eigenfunction of H with eigenvalue λ. Then $A^*\varphi$ is an eigenfunction with eigenvalue $\lambda + 2$, and $A\varphi$ is an eigenfunction (as long as $\lambda \neq 1$) with eigenvalue $\lambda - 2$.*

The proof is a simple computation involving the commutation identities we already computed. Of course we need also to observe that $A^*\varphi$ and $A\varphi$ are not identically zero, but we have already seen that $A\varphi = 0$ means $\lambda = 1$, and the solutions to $A^*\varphi = 0$ are $ce^{x^2/2}$ which are ruled out by growth conditions (we need at least $\int |\varphi(x)|^2\, dx < \infty$ for spectral theory to work). Now since $AH - HA = 2A$ we have $AH\varphi - HA\varphi = 2A\varphi$, hence

$$H(A\varphi) = AH\varphi - 2A\varphi = (\lambda - 2)A\varphi$$

and similarly

$$A^*H\varphi - HA^*\varphi = -2A^*\varphi,$$

hence

$$H(A^*\varphi) = A^*H\varphi + 2A^*\varphi = (\lambda + 2)A^*\varphi,$$

proving the lemma.

So the creation operator boosts the eigenvalue by 2 and the annihilation operator decreases it by 2. We immediately obtain an infinite ladder of eigenfunctions with eigenvalues $1, 3, 5, \ldots$ by applying powers of the creation operators to the groundstate. We write

$$h_n(x) = c_n(A^*)^n e^{-x^2/2}$$

where the positive constants c_n are chosen so that $\int h_n(x)^2\, dx = 1$ (in fact $c_n = \pi^{-1/4}(2^n n!)^{-1/2}$, but we will not use this formula). Then $Hh_n = (2n + 1)h_n$ by the lemma. We claim that there are no other eigenvalues, and each positive odd integer has multiplicity one (the space of eigenfunctions is one-dimensional, consisting of multiples of h_n). The reasoning goes as follows: if we start with any λ not a positive odd integer, then by applying a high enough power of the annihilation operator we would end up with an eigenvalue less than 1 (note that we never pass through the eigenvalue 1), which contradicts our observation that 1 is the bottom of the spectrum. Similarly, the fact that the eigenspace of 1 has multiplicity one is inherited inductively for the eigenspaces of $2n + 1$ because the annihilation operator is a one-to-one mapping down the ladder (it only fails to be one-to-one on the eigenspace of 1, which it does annihilate).

The functions $h_n(x)$ are called the *Hermite functions*. Explicitly

$$h_n(x) = c_n \left(\frac{d}{dx} - x \right)^n e^{-x^2/2},$$

and it is easy to see that $h_n(x) = c_n H_n(x) e^{-x^2/2}$ where $H_n(x)$ is a polynomial of degree n (in fact H_n is even or odd depending on the parity of n), called the *Hermite polynomial*. It is clear from this formula that $h_n \in \mathcal{S}$.

Now the Hermite functions form an orthonormal system (this is a general property of eigenfunctions of self-adjoint operators), namely $\langle h_n, h_m \rangle = \int h_n(x) h_m(x)\, dx = 0$ if $n \neq m$ (and $= 1$ for $n = m$). Note that since $h_n(x)$ are real-valued functions, we can omit complex conjugates. This follows from the identity

$$(2n + 1)\langle h_n, h_m \rangle = \langle H h_n, h_m \rangle = \langle h_n, H h_m \rangle = (2m + 1)\langle h_n, h_m \rangle$$

which is impossible for $n \neq m$ unless $\langle h_n, h_m \rangle = 0$. They are also a complete system, so any function f (with $\int |f|^2\, dx < \infty$) has an expansion

$$f = \sum_{n=0}^{\infty} \langle f, h_n \rangle h_n$$

called the *Hermite expansion* of f. The coefficients $\langle f, h_n \rangle$ satisfy the Parseval identity

$$\sum_{n=0}^{\infty} |\langle f, h_n \rangle|^2 = \int |f|^2\, dx.$$

The expansion is very well suited to the spaces $\mathcal{S}$ and $\mathcal{S}'$. We have $f \in \mathcal{S}$ if and only if the coefficients are rapidly decreasing, $|\langle f, h_n \rangle| \leq c_N (1+n)^{-N}$ all N. Notice that $\langle f, h_n \rangle$ is well defined for $f \in \mathcal{S}'$. In that case we have $|\langle f, h_n \rangle| \leq c(1 + n)^N$ for some N, and conversely $\sum a_n h_n$ represents a tempered distribution if the coefficients satisfy such a polynomial bound.

Now it is not hard to see that the Hermite functions are eigenfunctions for the Fourier transform. Indeed, we only have to check (from the pingpong table) the behavior of the creation and annihilation operators on the Fourier transform side, namely

$$\mathcal{F} A^* \varphi = i A^* \mathcal{F} \varphi$$

and

$$\mathcal{F}A\varphi = -iA\mathcal{F}\varphi.$$

From the first of these we find

$$
\begin{aligned}
\mathcal{F}h_n &= c_n \mathcal{F}(A^*)^n e^{-x^2/2} \\
&= c_n(i)^n (A^*)^n \mathcal{F} e^{-x^2}/2 \\
&= i^n \sqrt{2\pi}\, c_n (A^*)^n e^{-x^2/2} = i^n \sqrt{2\pi}\, h_n
\end{aligned}
$$

so h_n is an eigenfunction with eigenvalue $i^n \sqrt{2\pi}$. We also observe that

$$\mathcal{F}H\varphi = \mathcal{F}(A^*A + 1)\varphi = (A^*A + 1)\mathcal{F}\varphi = H\mathcal{F}\varphi$$

so H commutes with the Fourier transform. It is a general principle that commuting operators have common eigenfunctions, which explains why the spectral theory of H and $\mathcal{F}$ are so closely related.

Note that we can interpret this fact as saying that the Hermite expansion *diagonalizes* the Fourier transform. If $f = \sum a_n h_n$ then $\mathcal{F}f = \sum i^n \sqrt{2\pi}\, a_n h_n$, so that $\mathcal{F}$ is represented as a diagonal matrix if we use the basis $\{h_n\}$.

We also have another representation for all the eigenfunctions of the Fourier transform; for example, $\mathcal{F}f = \sqrt{2\pi}f$ if and only if f has a Hermite expansion $f = \sum a_{4n} h_{4n}$ containing only Hermite functions of order $\equiv 0$ mod 4. Again, this is not a very illuminating condition. For example, the Poisson summation formula implies that the distribution $f = \sum_{k=-\infty}^{\infty} \delta(x - \sqrt{2\pi}k)$ satisfies $\mathcal{F}f = \sqrt{2\pi}f$. I do not know how to compute the Hermite expansion of this distribution.

Using further properties of Hermite functions, it is possible to find explicit solutions to various differential equations involving H, such as the Schrödinger equation

$$\frac{\partial}{\partial t}u = ikHu$$

(and the n-dimensional analog) which is important in quantum mechanics.

7.7 Radial Fourier transforms and Bessel functions

The Fourier transform of a radial function is again a radial function. We observed this before indirectly, because the Fourier transform commutes with rotations, and radial functions are characterized by invariance under rotation. In this section we give a more direct explanation, including a formula for the radial Fourier transform. The formula involves a class of special functions called *Bessel functions*. We begin, as usual, with the one-dimensional case, where the result is very banal. In one-dimension, a radial function is just an even function, $f(x) = f(-x)$. The Fourier transform of an even function is also even, and is given by the Fourier cosine formula:

$$
\begin{aligned}
\hat{f}(\xi) &= \int_{-\infty}^{\infty} e^{ix\xi} f(x)\, dx \\
&= \int_{-\infty}^{\infty} (\cos x\xi + i \sin x\xi) f(x)\, dx \\
&= 2 \int_{0}^{\infty} f(x) \cos x\xi\, dx,
\end{aligned}
$$

the sine term vanishing because it is the integral of an odd function.

There is a similar formula for $n = 3$, but we have to work harder to derive it. We work in a spherical coordinate system. If f is radial we write $f(x) = f(|x|)$ by abuse of notation. Suppose we want to compute $\hat{f}(0, 0, R)$. Then we take the z-axis as the central axis and the spherical coordinates are

$$
\begin{aligned}
x &= r \sin \phi \cos \theta \\
y &= r \sin \phi \sin \theta \\
z &= r \cos \phi
\end{aligned}
$$

with $0 \le r < \infty, 0 \le \theta \le 2\pi, 0 \le \phi \le \pi$ and the element of integration is $dx\, dy\, dz = r^2 \sin \phi\, dr\, d\phi\, d\theta$ ($r^2 \sin \phi$ is the determinant of the Jacobian matrix $\partial(x, y, z)/\partial(r, \phi, \theta)$, a computation you should do if you have not seen it before). The Fourier transform formula is then

$$
\hat{f}(0, 0, R) = \int_{0}^{2\pi} \int_{0}^{\pi} \int_{0}^{\infty} e^{iRr \cos \phi} f(r) r^2 \sin \phi\, dr\, d\phi\, d\theta.
$$

Now the θ-integral produces a factor of 2π since nothing depends on θ. The ϕ-integral can also be done explicitly, since

$$
\begin{aligned}
\int_{0}^{\pi} e^{iRr \cos \phi} \sin \phi\, d\phi &= \left. \frac{-e^{iRr \cos \phi}}{iRr} \right|_{0}^{\pi} \\
&= \frac{e^{iRr} - e^{-iRr}}{iRr} = \frac{2 \sin rR}{rR}.
\end{aligned}
$$

Thus we have altogether

$$
\hat{f}(0, 0, R) = 4\pi \int_{0}^{\infty} \frac{\sin rR}{rR} f(r) r^2 dr.
$$

Since $\hat{f}$ is radial, this is the formula for $\hat{f}(R)$ (or, if you prefer, given any ξ with $|\xi| = R$, set up a spherical coordinate system with principle axis in the direction of ξ, and the computation is identical).

Superficially, the formula for $n = 3$ resembles the formula for $n = 1$, with the cosine replaced by the sine. But the appearance of rR in the denominator has some implications concerning the decay of Fourier transforms of radial functions. For example, suppose f is radial and integrable. We can almost conclude that $\hat{f}(R)$ decays faster than R^{-1} as $R \to \infty$. I say "almost" because we have to assume something about the behavior of f near zero; it suffices to have f continuous at zero, or even just bounded in a neighborhood of zero. The fact that f is integrable is equivalent to the finiteness of $\int |f(r)|r^2\, dr$. This certainly implies $\int_1^\infty |f(r)|r\, dr < \infty$, and the assumption about the boundedness of f near zero implies $\int_0^1 |f(r)|r\, dr < \infty$, so

$$\int_0^\infty |f(r)|r\, dr < \infty.$$

Then the Riemann-Lebesgue lemma for the Fourier sine transform of the one-dimensional function $|f(r)|r$ implies

$$\lim_{R \to \infty} \int_0^\infty \sin rR f(r)r\, dr = 0.$$

Since $\hat{f}(R) = 4\pi R^{-1} \int_0^\infty \sin rR f(r)r\, dr$ we obtain the desired conclusion. Of course this is much stronger than the conclusion of the 3-dimensional Riemann-Lebesgue lemma. The moral of the story is that Fourier transforms of radial functions are unusually well behaved.

So what is the story for other dimensions? $n = 2$ will illustrate the difficulties. If we use polar coordinates

$$x = r \cos \theta$$
$$y = r \sin \theta$$

for $0 \le r < \infty, 0 \le \theta \le 2\pi$, the element of integration is $dx\, dy = r\, dr\, d\theta$, and our computation as before becomes

$$\hat{f}(R, 0) = \int_0^{2\pi} \int_0^\infty e^{irR \cos \theta} f(r)r\, dr\, d\theta.$$

This time there is no friendly $\sin \theta$ factor, so we cannot do the θ-integration in terms of elementary functions. It turns out that $\int_0^{2\pi} e^{is \cos \theta} d\theta$ is a new kind of special function, a Bessel function of order zero. As the result of historical accident, the exact notation is

$$J_0(s) = \frac{1}{2\pi} \int_0^{2\pi} e^{is \cos \theta} d\theta.$$

Thus the 2-dimensional radial Fourier transform is given by

$$\hat{f}(R) = 2\pi \int_0^\infty J_0(rR) f(r) r \, dr.$$

This would fit the pattern of the other two cases if we could convince ourselves that J_0 behaves something like a cosine or sine times a power.

Now there are whole books devoted to the properties of J_0 and its cousins, the other Bessel functions, and precise numerical computations are available. Here I will only give a few salient facts. First we observe that by substituting the power series for the exponential we can integrate term by term, using the elementary fact

$$\frac{1}{2\pi} \int_0^{2\pi} \cos^{2k} \theta \, d\theta = \frac{(2k)!}{2^{2k}(k!)^2}$$

to obtain

$$
\begin{aligned}
J_0(s) &= \frac{1}{2\pi} \int_0^{2\pi} \sum_{k=0}^\infty \frac{(is \cos \theta)^k}{k!} \, d\theta \\
&= \sum_{k=0}^\infty \frac{(-1)^k s^{2k}}{(2k)!} \frac{1}{2\pi} \int_0^{2\pi} \cos^{2k} \theta \, d\theta \\
&= \sum_{k=0}^\infty \frac{(-1)^k s^{2k}}{2^{2k}(k!)^2}
\end{aligned}
$$

(the integrals of the odd powers are zero by cancelation). This power series converges everywhere, showing that J_0 is an even entire analytic function. Also $J_0(0) = 1$. Notice that these properties are shared by $\cos s$ and $\sin s/s$, which are the analogous functions for $n = 1$ and $n = 3$.

The behavior of $J_0(s)$ as $s \to \infty$ is more difficult to discern. First let's make a change of variable to reduce the integral defining J_0 to a Fourier transform: $t = \cos \theta$. We obtain

$$J_0(s) = \frac{1}{\pi} \int_{-1}^1 e^{ist} (1 - t^2)^{-1/2} dt.$$

Thus J_0 is the 1-dimensional Fourier transform of the function we can write $(1 - t^2)_+^{-1/2}$ (the $+$ subscript means we set the function equal to zero when $1 - t^2$ becomes negative, i.e., outside $|t| < 1$). This function has compact support and is integrable, even though it has singularities at $t = \pm 1$. Note that the Paley-Wiener theorems imply that J_0 is an entire function, as we observed already, and in fact it has exponential type 1. Also the Riemann-Lebesgue lemma implies J_0 vanishes at infinity, but we are seeking much more precise information. Now we claim that only the part of the integral near the singularities $t = \pm 1$ contributes to the interesting behavior of J_0 near infinity. If we were to multiply $(1 - t^2)_+^{-1/2}$ by a cut-off function vanishing in a neighborhood of the singularities $t = \pm 1$, the result would be a function in $\mathcal{D}$, whose Fourier transform is in $\mathcal{S}$ hence decays faster than any polynomial rate. Now we can factor the function $(1 - t^2)_+^{-1/2} = (1 - t)_+^{-1/2}(1 + t)_+^{-1/2}$. In a neighborhood of $t = +1$, the function $(1 + t)_+^{-1/2}$ is C^∞, and to first approximation is just the constant $2^{-1/2}$. Similarly, in a neighborhood of $t = -1$, the function $(1 - t)_+^{-1/2}$ is C^∞ and is approximately $2^{-1/2}$. Thus, to first approximation, we should have

$$
\begin{aligned}
J_0(s) &\approx \frac{1}{\sqrt{2\pi}} \int_{-\infty}^{\infty} e^{ist}(1 - t)_+^{-1/2} dt \\
&\quad + \frac{1}{\sqrt{2\pi}} \int_{-\infty}^{\infty} e^{ist}(1 + t)_+^{-1/2} dt \\
&= \frac{1}{\sqrt{2\pi}} \int_0^{\infty} e^{is} e^{-isx} x^{-1/2} dx \\
&\quad + \frac{1}{\sqrt{2\pi}} \int_0^{\infty} e^{-is} e^{isx} x^{-1/2} dx
\end{aligned}
$$

(we made the change of variable $x = 1 - t$ in the first integral and $x = 1 + t$ in the second). However, by a modification of the argument in example 5 of 4.2 we have

$$
\int_0^{\infty} e^{isx} x^{-1/2} dx = \begin{cases} i\sqrt{\pi}\, e^{-i\pi/4} s^{-1/2} & s > 0 \\ -i\sqrt{\pi}\, e^{i\pi/4} |s|^{-1/2} & s < 0. \end{cases}
$$

Thus for $s > 0$ large, we have

$$
\begin{aligned}
J_0(s) &\approx \frac{1}{\sqrt{2}\sqrt{\pi}}(ie^{is}(-e^{i\pi/4}) + ie^{-is}e^{-i\pi/4}) s^{-1/2} \\
&\approx \sqrt{\frac{2}{\pi}}\left(\frac{e^{i(s+\pi/4)} - e^{-i(s+\pi/4)}}{2i}\right) s^{-1/2} \\
&\approx \sqrt{\frac{2}{\pi}} s^{-1/2} \sin(s + \pi/4).
\end{aligned}
$$

This is exactly what we were looking for: the product of a power and a sine function. Of course our approximate calculation only suggests this answer, but a more careful analysis of the error shows that this is correct; in fact the error is of order $s^{-3/2}$. In fact, this is just the first term of an asymptotic expansion (as $s \to \infty$) involving powers of s and translated sines.

To unify the three special cases $n = 1, 2, 3$ we have considered, we need to give the definition of the Bessel function J_α of arbitrary (at least $\alpha > -\frac{1}{2}$) order:

$$J_\alpha(s) = \gamma_\alpha s^\alpha \int_{-1}^{1} (1 - t^2)^{\alpha - 1/2} e^{ist} dt$$

(the constant γ_α is given by $\gamma_\alpha = 1/\sqrt{\pi} 2^\alpha \Gamma(\alpha + \frac{1}{2})$). Note that when $\alpha = \frac{1}{2}$ it is easy to evaluate

$$
\begin{aligned}
J_{1/2}(s) &= \frac{1}{\sqrt{2\pi}} s^{1/2} \int_{-1}^{1} e^{ist} dt \\
&= \sqrt{\frac{2}{\pi}} s^{-1/2} \sin s.
\end{aligned}
$$

Using integration by parts we can prove the following recursion relations for Bessel functions of different orders:

$$
\begin{aligned}
\frac{d}{ds}(s^\alpha J_\alpha(s)) &= s^\alpha J_{\alpha-1}(s) \\
\frac{d}{ds}(s^{-\alpha} J_\alpha(s)) &= -s^{-\alpha} J_{\alpha+1}(s).
\end{aligned}
$$

Using these we can compute

$$J_{-1/2}(s) = \sqrt{\frac{2}{\pi}} s^{-1/2} \cos s,$$

and more generally, if $\alpha = k + \frac{1}{2}, k$ an integer, then $J_{k+\frac{1}{2}}$ is expressible as a finite sum of powers and sines and cosines. The function $s^{-\alpha} J_\alpha(s)$ is clearly an entire analytic function. Also, the asymptotic behavior of $J_\alpha(s)$ as $s \to +\infty$ is given by

$$J_\alpha(s) \approx \sqrt{\frac{2}{\pi}} s^{-1/2} \cos \left(s - \frac{\alpha\pi}{2} - \frac{\pi}{4} \right).$$

Now each of the formulas for the radial Fourier transform for $n = 1, 2, 3$ can be written

$$\hat{f}(R) = c_n \int_0^\infty \frac{J_{\frac{n-2}{2}}(rR)}{(rR)^{\frac{n-2}{2}}} f(r) r^{n-1} dr$$

and in fact the same is true for general n. The general form of the decay rate for radial Fourier transforms is the following: if f is a radial function that is integrable, and bounded near the origin, then $\hat{f}(R)$ decays faster than $R^{-(\frac{n-1}{2})}$.

Many radial Fourier transforms can be computed explicitly, using properties of Bessel functions. For example, the Fourier transform of $(1 - |x|^2)_+^\alpha$ in $\mathbb{R}^n$ is equal to

$$c(n, \alpha) \frac{J_{\frac{n}{2}+\alpha}(R)}{R^{\frac{n}{2}+\alpha}}$$

for the appropriate constants $c(n, \alpha)$.

7.8 Haar functions and wavelets

Fourier analysis shows us how to expand an arbitrary function in terms of the simple basic functions $e^{ix \cdot \xi}$. These functions are precisely determined in frequency, but have no localization in space. We have seen that the Heisenberg uncertainty principle prohibits a function from simultaneously being very localized in space and frequency. The Gabor transform and its inversion formula gives a way of generating all functions out of a single prototype, the Gaussian, which is translated in both space and frequency. The Gaussian is reasonably localized on both accounts, and we cannot do any better from that point of view. Wavelet theory is a different attempt to expand a general function in terms of simple building blocks that are reasonably localized in space and frequency. The difference is that we do not translate on the frequency side; instead we take both translates and dilates of a single wavelet.

The simplest example of such a wavelet is the Haar function, defined as

$$\psi(x) = \begin{cases} 1 & \text{if} \quad 0 < x \leq \frac{1}{2} \\ -1 & \text{if} \quad \frac{1}{2} < x \leq 1 \\ 0 & \text{otherwise.} \end{cases}$$

The support of ψ is $[0, 1]$. We dilate ψ by powers of 2, so $\psi(2^j x)$ is supported on $[0, 2^{-j}]$ (j is an integer, not necessarily positive) and we translate the dilate by 2^{-j} times an integer, to obtain

$$\psi(2^j(x - 2^{-j}k)) = \psi(2^j x - k).$$

This function is supported on $[2^{-j}k, 2^{-j}(k + 1)]$. It is also convenient to normalize this function by multiplying by $2^{j/2}$. The family of functions $2^{j/2}\psi(2^j x - k) = \psi_{j,k}(x)$ is an orthonormal family, meaning $\int \psi_{j,k}(x)\psi_{j',k'}(x)\, dx = 0$ unless $j = j'$ and $k = k'$, in which case the

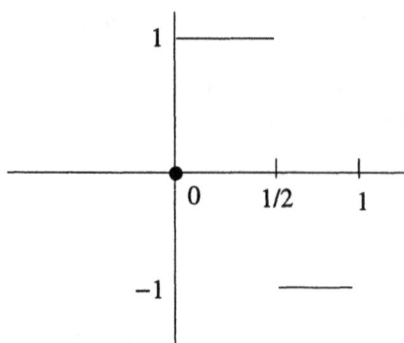

Figure 7.2

integral is one (the condition $\int \psi_{j,k}(x)^2\,dx = 1$ is satisfied exactly because of the normalization factor $2^{j/2}$ we chose). We can understand the orthogonality condition very simply. If $j = j'$ then the functions $\psi_{j,k}$ and $\psi_{j,k'}$ have disjoint support if $k \neq k'$, so their product is identically zero. If $j \neq j'$ then $\psi_{j,k}$ and $\psi_{j',k'}$ may have overlapping support, but we claim that the integral of their product must still be zero, because of cancellation. Say $j > j'$. Then $\psi_{j,k}$ changes sign on neighboring intervals of length 2^{-j}. But $\psi_{j',k'}$ is constant on intervals of length $2^{-j'}$, in particular on the support of $\psi_{j,k}$, so

$$\int \psi_{j,k}(x)\psi_{j',k'}(x)\,dx = \text{const} \int \psi_{j,k}(x)\,dx = 0.$$

Because we have an orthonormal family, we can attempt to expand an arbitrary function according to the recipe

$$f = \sum_{j=-\infty}^{\infty} \sum_{k=-\infty}^{\infty} \langle f, \psi_{j,k} \rangle \psi_{j,k}.$$

We refer to this as the *Haar series expansion* of f. The system is called *complete* if the expansion is valid. Of course this is something of an oversimplification for two reasons:

1. We cannot expect to represent a completely arbitrary function, the usual minimal restriction being $\int |f(x)|^2\,dx < \infty$.

2. The convergence of the infinite series representation may not be valid in the pointwise sense and may not have certain intuitively plausible properties.

Since both these problems already occur with ordinary Fourier series, we should not be surprised to meet them again here.

We claim in the case of Haar functions that we do have completeness. Before demonstrating this, we should point out one paradoxical consequence of this claim. Each of the functions $\psi_{j,k}$ has total integral zero, $\int_{-\infty}^{\infty} \psi_{j,k}(x)\,dx = 0$, so any linear combination will also have zero integral. If we start out with a function f whose total integral is nonzero, how can we expect to write it as a series in functions with zero integral?

The resolution of the paradox rests on the fact that an infinite series $\sum_{j,k} c_{jk}\psi_{j,k}(x)$ is not quite the same as a finite linear combination. The argument that

$$\int \sum_{j,k} c_{jk}\psi_{j,k}(x)\,dx = \sum_{j,k} c_{jk} \int \psi_{j,k}(x)\,dx = \sum_{j,k} c_{j,k} \cdot 0 = 0$$

requires the interchange of an integral and an infinite sum, and such an interchange is not always valid. In particular, the convergence of the Haar series expansion does not allow term-by-term integration.

We can explain the existence of Haar series expansions if we accept the following general criterion for completeness of an orthonormal system: the system is complete if and only if any function whose expansion is identically zero (i.e., $\langle f, \psi_{j,k} \rangle = 0$ for all j and k) must be the zero function (in the sense discussed in section 2.1). We can verify this criterion rather easily in our case, at least under the assumption that f is integrable (this is not quite the correct assumption, which should be that $|f|^2$ is integrable).

So suppose f is integrable and $\langle f, \psi_{j,k} \rangle = 0$ for all j and k. First we claim $\int_0^{1/2} f(x)\,dx = 0$. Why? Since $\langle f, \psi_{0,0} \rangle = 0$ we know

$$\int_0^{1/2} f(x)\,dx = \int_{1/2}^1 f(x)\,dx,$$

so

$$\int_0^1 f(x)\,dx = 2\int_0^{1/2} f(x)\,dx.$$

But then since $\langle f, \psi_{-1,0} \rangle = 0$ we know

$$\int_0^1 f(x)\,dx = \int_1^2 f(x)\,dx$$

so

$$\int_0^2 f(x)\,dx = 4\int_0^{1/2} f(x)\,dx.$$

By repeating this argument we obtain

$$\int_0^{2^k} f(x)\,dx = 2^{k+1}\int_0^{1/2} f(x)\,dx$$

and this would contradict the integrability condition unless $\int_0^{1/2} f(x)\,dx = 0$. However, there is nothing special about the interval $[0,\frac{1}{2}]$. By the same sort of argument we can show $\int_I f(x)\,dx$ for any interval I of the form $[2^{-j}k, 2^{-j}(k+1)]$ for any j and k, and by additivity of the integral, for any I of the form $[2^{-j}k, 2^{-j}m]$ for any integers j,k,m. But an arbitrary interval can be approximated by intervals of this form, so we obtain $\int_I f(x)\,dx = 0$ for any interval I. This implies that f is the zero function in the correct sense.

There is a minor variation of the Haar series expansion that has a nicer interpretation and does not suffer from the total integral paradox. To explain it, we need to introduce another function, denoted φ, which is simply the characteristic function of the interval $[0,1]$. We define $\varphi_{j,k}$ by dilation and translation as before: $\varphi_{j,k}(x) = 2^{j/2}\varphi(2^j x - k)$. We no longer claim that $\varphi_{j,k}$ is an orthonormal system, since we have no cancellation properties. However, all the functions $\varphi_{j,k}$ with one fixed value of j are orthonormal, because they have disjoint supports. Also, if $j' \geq j$, then the previous argument shows $\int \varphi_{j,k}(x)\psi_{j',k'}(x)\,dx = 0$. Thus, in particular, the system consisting of $\psi_{0,k}$ for all k and $\psi_{j,k}$ for all k and just $j \geq 0$ is orthonormal and a variant of the previous argument shows that it is also complete. The associated expansion is

$$f = \sum_{k=-\infty}^{\infty} \langle f, \varphi_{0,k}\rangle\varphi_{0,k} + \sum_{j=0}^{\infty}\sum_{k=-\infty}^{\infty} \langle f, \psi_{j,k}\rangle\psi_{j,k}.$$

Now here is the interpretation of this expansion. The first series,

$$\sum_{k=-\infty}^{\infty} \langle f, \varphi_{0,k}\rangle\varphi_{0,k}$$

is just a coarse approximation to f by a step function with steps of length one. Each of the subsequent series

$$\sum_{k=-\infty}^{\infty} \langle f, \psi_{0,k}\rangle\psi_{0,k}$$

for $j = 0,1,2,\ldots$ adds finer and finer detail on the scale of 2^{-j}. One of the advantages of this expansion is its excellent space localization. Each of the

Haar functions is supported on a small interval; the larger j, the smaller the interval. If the function f being expanded vanishes on an interval I, then the coefficients of f will be zero for all $\psi_{j,k}$ whose support lies in I.

A grave disadvantage of the Haar series expansions is that the Haar functions are discontinuous. Thus, no matter how smooth the function f may be, the approximations obtained by taking a finite number of terms of the expansion will be discontinuous. This defect is remedied in the smooth wavelet expansions we will mention shortly.

Another closely related defect in the Haar series expansion is the lack of localization in frequency. We can compute directly the Fourier transform of φ:

$$
\begin{aligned}
\hat{\varphi}(\xi) &= \int_0^1 e^{ix\xi}\,dx = \frac{e^{i\xi}-1}{i\xi} \\
&= 2e^{i\xi/2}\frac{\sin\xi/2}{\xi}.
\end{aligned}
$$

Aside from the oscillatory factors, this decays only like $|\xi|^{-1}$, which means it is not integrable (of course we should have known this in advance: a discontinuous function cannot have an integrable Fourier transform). The behavior of $\hat{\psi}$ is similar. In fact, since

$$
\psi(x) = \varphi(2x) - \varphi(2x-1)
$$

we have

$$
\hat{\psi}(\xi) = \frac{1}{2}(1 - e^{i\xi/2})\hat{\varphi}(\xi/2)
$$

which has the same $|\xi|^{-1}$ decay rate.

The Haar system has been around for more than half a century, but has not been used extensively because of the defects we have mentioned. In the last five years, a number of related expansions, called wavelet expansions, have been introduced. Generally speaking, there are two functions, φ, called the *scaling function*, and ψ, called the *wavelet*, so that the identical formulas hold. The particular system I will describe, called *Daubechies wavelets*, are compactly supported and differentiable a finite number of times. Other wavelet systems have somewhat different support and smoothness properties. The key to generating wavelets is the observation that the Haar system possesses a kind of self-similarity. Not only is ψ expressible as

$$
\psi(x) = \varphi(2x) - \varphi(2x-1),
$$

a linear combinations of translates and dilates of φ, but so is φ itself:

$$
\varphi(x) = \varphi(2x) + \varphi(2x-1).
$$

This identity actually characterizes φ, up to a constant multiple. It is referred to as a *scaling identity* or a *dilation equation*. Because of the factor of 2 on the right side, it says that φ on a larger scale is essentially the same as φ on a smaller scale. The scaling function for the Daubechies wavelets satisfies an analogous but more complicated scaling identity,

$$\varphi(x) = \sum_{k=0}^{N} a_k \varphi(2x - k).$$

The coefficients a_k must be chosen with extreme care, and we will not be able to say more about them here. The number of terms, $N + 1$, has to be taken fairly large to get smoother wavelets (approximately 5 times the number of derivatives desired).

The scaling identity determines φ, up to a constant multiple. If we restrict φ to the integers, then the scaling identity becomes an eigenvalue equation for a finite matrix, which can be solved by linear algebra. Once we know φ on the integers, we can obtain the values of φ on the half-integers from the scaling identity (if $x = m/2$ on the left side, then the values $2x - k = m - k$ on the right are all integers). Proceeding inductively, we can obtain the values of φ at dyadic rationals $m2^{-j}$, and by continuity this determines $\varphi(x)$ for all x. Then the wavelet ψ is determined from φ by the identity

$$\psi(x) = \sum_{k=0}^{N} (-1)^k a_{N-k} \varphi(2x - k).$$

These scaling functions and wavelets should be thought of as new types of special functions. They are not given by any simple formulas in terms of more elementary functions, and their graphs do not look familiar (see figures 7.3 and 7.4).

However, there exist very efficient algorithms for computing the coefficients of wavelet expansions, and there are many practical purposes for which wavelet expansions are now being used.

It turns out that we can compute the Fourier transforms of scaling functions and wavelets rather easily. This is because the scaling identity implies

$$\hat{\varphi}(\xi) = \frac{1}{2} \sum_{k=0}^{N} a_k e^{ik\xi} \hat{\varphi}(\xi/2)$$

which we can write

$$\hat{\varphi}(\xi) = p(\xi/2)\hat{\varphi}(\xi/2)$$

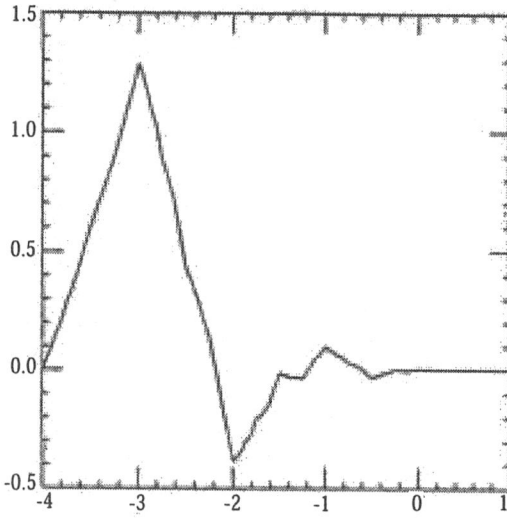

Figure 7.3.   The graph of the scaling function φ, courtesy of David Aronstein.

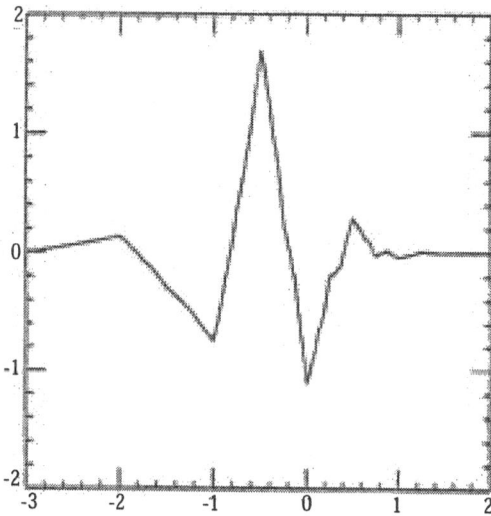

Figure 7.4.   The graph of the scaling function φ, courtesy of David Aronstein.

where

$$p(\xi) = \frac{1}{2} \sum_{k=0}^{N} a_k e^{ik\xi}.$$

By iterating this identity we obtain

$$\hat{\varphi}(\xi) = \left(\prod_{j=1}^{m} p(\xi/2^j) \right) \hat{\varphi}(\xi/2^m).$$

It is convenient to normalize φ so that $\hat{\varphi}(0) = 1$. Then since $\xi/2^m \to 0$ as $m \to \infty$ we obtain the infinite product representation

$$\hat{\varphi}(\xi) = \prod_{j=1}^{\infty} p(\xi/2^j).$$

(Note: the coefficients are always chosen to satisfy $\frac{1}{2} \sum_{k=0}^{N} a_k = 1$ so $p(0) = 1$, which makes the infinite product converge—in fact it converges so rapidly that only 10 or 20 terms are needed to compute $\hat{\varphi}(\xi)$ for small values of ξ.) Then $\hat{\psi}$ is expressible in terms of $\hat{\varphi}$ as

$$\hat{\psi}(\xi) = q(\xi/2)\hat{\varphi}(\xi)$$

where

$$q(\xi) = \frac{1}{2} \sum_{k=0}^{N} (-1)^k a_{N-k} e^{ik\xi}.$$

It is interesting to compare the infinite product form of $\hat{\varphi}$ and the direct computation of $\hat{\varphi}$ in the case of the Haar functions. Here $p(\xi) = \frac{1}{2}(1+e^{i\xi}) = e^{i\xi/2} \cos \xi/2$, so

$$\hat{\varphi}(\xi) = \prod_{j=2}^{\infty} e^{i\xi/2^j} \cos \xi/2^j = e^{i\xi/2} \prod_{j=2}^{\infty} \cos \xi/2^j.$$

This is the same result as before because

$$\prod_{j=1}^{\infty} \cos x/2^k = \sin x/x,$$

an identity known to Euler, and in special cases, Francois Viete in the 1590s.

7.9   Problems

1. Show that

$$\lim_{t \to 0} \int_a^b f(x) \sin(tg(x))\, dx = 0$$

   if f is integrable and g is C^1 on $[a, b]$ with nonvanishing derivative.

2. Show that an estimate

$$|\hat{f}(\xi)| \leq c\, \|f\|_1\, h(\xi)$$

   cannot hold for all integrable functions, where $h(\xi)$ is any fixed function vanishing at infinity, by considering the family of functions $f(x)e^{i\eta \cdot x}$ for fixed f and η varying.

3. Show that an estimate of the form

$$\|\hat{f}\|_q \leq c\|f\|_p$$

   for all f with $\|f\|_p < \infty$ cannot hold unless $\frac{1}{p} + \frac{1}{q} = 1$, by considering the dilates of a fixed function.

4. Let $f(x) \sim \sum c_n e^{inx}$ be the Fourier series of a periodic function with $\int_0^{2\pi} |f(x)|\, dx < \infty$. Show that

$$\lim_{n \to \pm\infty} c_n = 0.$$

5. Suppose $|f(x)| \leq ce^{-a|x|}$. Prove that $\hat{f}$ extends to an analytic function on the region $|\operatorname{Im} \zeta| < a$.

6. Show that the distributions

$$\mathcal{F}^{-1}\left(\cos t\sqrt{m^2 + |\xi|^2}\right)$$

   and

$$\mathcal{F}^{-1}\left(\frac{\sin t\sqrt{m^2 + |\xi|^2}}{\sqrt{m^2 + |\xi|^2}}\right)$$

   are supported in $|x| \leq |t|$. Use this to prove a finite speed of propagation for solutions of the Klein–Gordon equation

$$\frac{\partial^2 u}{\partial t^2} = \Delta_x u - m^2 u.$$

7. Let f be an integrable function supported in the half-plane $ax+by \geq 0$ in $\mathbb{R}^2$. What can you say about analyticity of the Fourier transform $\hat{f}$? What can you say if f is supported in the first quadrant ($x \geq 0$ and $y \geq 0$)?

8. What can you say about the product of two entire functions of exponential type? How does this relate to the support of the convolution of two functions of compact support?

9. Characterize the Fourier transforms of test functions on $\mathbb{R}^1$ supported in an arbitrary interval $[a, b]$.

10. Characterize the Fourier transforms of test functions on $\mathbb{R}^2$ supported in the rectangle $|x| \leq a, |y| \leq b$.

11. Let $f \in \mathcal{E}'(\mathbb{R}^1)$. Show that $\hat{f}(\zeta) = 0$ if and only if $g(x) = e^{i\zeta x}$ is a solution of the convolution equation $f * g = 0$. Show that ζ has multiplicity at least m (as a zero of $\hat{f}$) if and only if $g(x) = x^{m-1}e^{i\zeta x}$ is a solution.

12. Show that if f and g are distributions of compact support and not identically zero then $f * g$ is not identically zero.

13. Evaluate $\sum_{m=-\infty}^{\infty}(1+tm^2)^{-1}$ using the Poisson summation formula.

14. Compute the Fourier transform of $\sum_{k=-\infty}^{\infty}\delta'(x-k)$. What does this say about $\sum_{k=-\infty}^{\infty}f'(k)$ for suitable functions f?

15. Let Γ be the equilateral lattice in $\mathbb{R}^2$ generated by vectors $(1,0)$ and $(1/2, \sqrt{3}/2)$. What is the dual lattice? Sketch both lattices.

16. What is the Fourier transform of $\sum_{k=-\infty}^{\infty}\delta(x - x_0 - k)$ for fixed x_0?

17. Compute

$$\sum_{k=1}^{\infty}\left(\frac{\sin tk}{k}\right)^2$$

using the Poisson summation formula.

18. Is a translate of a positive definite function necessarily positive definite? What about a dilate?

19. Show that $e^{-|x|}$ is positive definite. (*Hint*: What is its Fourier transform?)

20. Show that the product of continuous positive definite functions is positive definite.

21. Show that if f and g are continuous positive definite functions on $\mathbb{R}^1$, $f(x)g(y)$ is positive definite on $\mathbb{R}^2$.

22. Show that $f * \tilde{f}$ is always positive definite, where $\tilde{f}(x) = \overline{f(-x)}$ and $f \in \mathcal{S}$.

23. Let μ be a probability measure on $[0, 2\pi)$. Characterize the coefficients of the Fourier series of μ in terms of positive definiteness.

24. Compute Var(ψ) when ψ is a Gaussian. Use this to check that the uncertainty principle inequality is an equality in this case.

25. Show that if a distribution and its Fourier transform both have compact support, then it is the zero distribution.

26. If ψ has compact support, obtain an upper bound for Var(ψ).

27. Let $\psi_t(x) = t^{-n/2}\psi(x/t)$, for a fixed function ψ. How does Var(ψ_t) depend on t? How does Var(ψ_t)Var($(2\pi)^{-n/2}\hat{\psi}_t$) depend on t?

28. Show that the Hermite polynomials satisfy $H_n(x) = (-1)^n e^{x^2} (d/dx)^n e^{-x^2}$.

29. Show that
$$H_n(x) = \sum_{k=0}^{[n/2]} \frac{(-1)^k n!}{k!(n-2k)!} (2x)^{n-2k}.$$

30. Prove the generating function identity
$$\sum_{n=0}^{\infty} \frac{H_n(x)}{n!} t^n = e^{2xt - t^2}.$$

31. Show that $H_n'(x) = 2nH_{n-1}(x)$.

32. Prove the recursion relation
$$H_{n+1}(x) - 2xH_n(x) + 2nH_{n-1}(x) = 0.$$

33. Compute the Fourier transform of the characteristic function of the ball $|x| \leq b$ in $\mathbb{R}^3$.

34. Prove the recursion relations
$$\frac{d}{ds}\left(s^\alpha J_\alpha(s)\right) = s^\alpha J_{\alpha-1}(s)$$
and
$$\frac{d}{ds}\left(s^{-\alpha} J_\alpha(s)\right) = -s^{-\alpha} J_{\alpha+1}(s).$$
Use this to compute $J_{3/2}$ explicitly.

35. Show that $J_\alpha(s)$ is a solution of Bessel's differential equation

$$f''(s) + \frac{1}{s}f'(s) + \left(1 - \frac{\alpha^2}{s^2}\right)f(s) = 0.$$

36. Show that

$$J_{n+\frac{1}{2}}(s) = (-1)^n \sqrt{\frac{2}{\pi}} s^{n+\frac{1}{2}} \left(\frac{1}{s}\frac{d}{ds}\right)^n \frac{\sin s}{s}$$

for n a positive integer.

37. Show that the integer translates of a fixed function f on $\mathbb{R}^1$ are orthonormal, i.e.,

$$\int_{-\infty}^\infty f(x-k)\overline{f(x-m)}\,dx = \begin{cases} 0 & k \neq m \\ 1 & k = m \end{cases}$$

if and only if $\hat{f}$ satisfies

$$\sum_{k=-\infty}^\infty |\hat{f}(\xi + 2\pi k)|^2 \equiv 1.$$

38. Show that the three families of functions $\varphi_{j,k}(x)\psi_{j,k'}(y), \psi_{j,k}(x)\varphi_{j,k'}(y)$ and $\psi_{j,k}(x)\psi_{j,k'}(y)$ for j, k, and k' varying over the integers, form a complete orthonormal system in $\mathbb{R}^2$, where ψ is the Haar function and φ the associated scaling function (the same for any wavelet and scaling function).

39. Show that a wavelet satisfies the vanishing moment conditions

$$\int_{-\infty}^\infty x^m \psi(x)\,dx = 0, \qquad m = 0, 1, \dots, M$$

provided

$$q(\xi) = \frac{1}{2}\sum_{k=0}^N (-1)^k a_{N-k} e^{ik\xi}$$

has a zero of order $M+1$ at $\xi = 0$.

40. Let V_j denote the linear span of the functions $\varphi_{j,k}(x)$ as k varies over the integers, where φ satisfies a scaling identity

$$\varphi(x) = \sum_{k=0}^N a_k \varphi(2x - k).$$

Show that $V_j \subseteq V_{j+1}$. Also show that $f(x) \in V_j$ if and only if $f(2x) \in V_{j+1}$.

41. Let φ and ψ be defined by $\hat{\varphi}(\xi) = \chi[-\pi, \pi](\xi)$ and

$$\hat{\psi}(\xi) = \chi[-2\pi, -\pi](\xi) + \chi[\pi, 2\pi](\xi).$$

Compute φ and ψ explicitly. Show that $\psi_{j,k}$ is a complete orthonormal system. (*Hint*: Work on the Fourier transform side.)

Chapter 8

Sobolev Theory and Microlocal Analysis

8.1 Sobolev inequalities

Throughout this work I have stressed the revolutionary quality of distribution theory: instead of asking "does this problem have any function solutions?," we ask "does this problem have any distribution solutions?" Nevertheless, there are times when you really want function solutions. We claim that even then distribution theory can be very useful. First we find the distribution solutions, and then we pose the question "when are the distribution solutions actually function solutions?" In order to answer such a question, we need a technique for showing that certain distributions are in fact functions. This technique is Sobolev theory.

Now in fact, Sobolev theory does not work miracles. We can never make the δ-distribution into a function. So in Sobolev theory, the hypotheses usually include the assumption that the distribution in question does come from a function, but we do not make any smoothness assumptions on the function. The conclusion will be that the function does have some smoothness, say it is C^k. This means that if the distribution is a solution of a differential equation of order less than or equal to k, in the distribution sense, then it satisfies the differential equation in the usual, pointwise sense. In this way, Sobolev theory allows you to obtain conventional solutions using distribution theory. Sobolev theory actual pre-dates distribution theory by more than a decade, so our placement of this topic is decidedly counter-historical.

The concepts of Sobolev theory involve a blend of smoothness and size measurements of functions. The size measurements involve the integrals of powers of the function $\int |f(x)|^p \, dx$ where p is a real parameter satisfying $p \geq 1$. We have already encountered the two most important special cases, $p = 1$ and 2. The finiteness of $\int |f(x)| \, dx$ is what we have called "integrability" of f. The quantity $\int |f(x)|^2 \, dx$ occurs in the Plancherel formula. If $|f(x)| > 1$ then $|f(x)|^p$ increases with p, so that as far as the singularities of f are concerned, the larger p is, the more information is contained in controlling the size of $\int |f(x)|^p \, dx$. But because the situation is reversed when $|f(x)| < 1$, there is no necessary relationship between the values of $\int |f(x)|^p \, dx$ for different values of p.

The standard terminology is to call $(\int |f(x)|^p \, dx)^{1/p}$ the L^p *norm* of f, written $||f||_p$. The conditions implied by the use of the word "norm" are as follows:

1. $||f||_p \geq 0$ with equality only for the zero function (positivity)

2. $||cf||_p = |c| \, ||f||_p$ (homogeneity)

3. $||f + g||_p \leq ||f||_p + ||g||_p$ (triangle inequality).

The first two conditions are evident (with $f = 0$ taken in the appropriate sense) but the proof of the triangle inequality is decidedly tricky (except when $p = 1$) and we will not discuss it. The case $p = \infty$, called the L^∞ norm, is defined separately by

$$||f||_\infty = \sup_x |f(x)|.$$

(**Exercise**: Verify conditions 1, 2, 3 for this norm.) The justification for this definition is the fact that

$$\lim_{p \to \infty} ||f||_p = ||f||_\infty,$$

which is valid under suitable restrictions so that both sides exist (a simple "counterexample" is $f \equiv 1$, for which $||f||_\infty = 1$ but $||f||_p = +\infty$ for all $p < \infty$). Suppose that f is continuous with compact support. Say f vanishes outside $|x| \leq R$ and $||f||_\infty = M$. Then

$$||f||_p = \left(\int |f(x)|^p \, dx \right)^{1/p} \leq M(cR^n)^{1/p}$$

where the constant is the volume of the unit ball in $\mathbb{R}^n$. Since $\lim_{p \to \infty} a^{1/p} = 1$ for any positive number a, we have $\lim_{p \to \infty} ||f||_p \leq M$. On the other hand, $||f||_\infty = M$ means that for every $\epsilon > 0$ there is an open set U on which $|f(x)| \geq M - \epsilon$; say U has volume A. Then

$$||f||_p \geq \left(\int_U |f(x)|^p \right)^{1/p} \geq (M - \epsilon)A^{1/p}$$

so $\lim_{p \to \infty} ||f||_p \geq M - \epsilon$. Since this is true for all $\epsilon > 0$, we have shown $\lim_{p \to \infty} ||f||_p = ||f||_\infty$.

Now the way we blend smoothness with L^p norm measurements of size is to consider L^p norms of derivatives. These are distributional derivatives, and a priori do not imply the existence of ordinary derivatives. Sobolev theory allows you to "trade in" a certain number of these L^p derivatives and get honest derivatives in return. For example, it will cost you exactly n derivatives in L^1 to get honest derivatives (n is the dimension of the space). We can state the result precisely as follows:

Theorem 8.1.1 $(L^1$ *Sobolev inequality*): *Let f be an integrable function on $\mathbb{R}^n$. Suppose the distributional derivatives $(\partial/\partial x)^\alpha f$ are also integrable functions for all $|\alpha| \leq n$. Then f is continuous and bounded, and*

$$||f||_\infty \leq c \sum_{|\alpha| \leq n} ||(\partial/\partial x)^\alpha f||_1 .$$

More generally, if $(\partial/\partial x)^\alpha f$ are integrable functions for $|\alpha| \leq n + k$, then f is C^k.

The terminology "Sobolev *inequality*" may seem strange. Notice there is an inequality in the middle of our stated theorem; this is the L^1 Sobolev inequality. It turns out that this is the key idea in the proof, and the remaining parts of the theorem follow from it in fairly routine fashion (once again we will omit most of this techical routine). Sometimes the term "Sobolev *embedding* theorem" is used. We will explain what this is about in the next section.

As usual, it is easiest to understand the case $n = 1$. Let us suppose, at first, that the function f is very well behaved, say even $f \in \mathcal{D}$, and let's see if we can prove the inequality. The fundamental theorem of the calculus allows us to write

$$f(x) = \int_{-\infty}^x f'(t)dt$$

(because f has compact support the lower endpoint is really finite and f vanishes there). Then we have

$$|f(x)| \leq \int_{-\infty}^x |f'(t)| \, dt \leq \int_{-\infty}^\infty |f'(t)| \, dt$$

and taking the supremum over x we obtain

$$||f||_\infty \le ||f'||_1.$$

This result looks somewhat better than the L^1 Sobolev inequality, but it was purchased by two strong hypotheses: differentiability and compact support. It is clearly nonsense without compact support (or at least vanishing at infinity). The constant function $f \equiv 1$ has $||f||_\infty = 1$ and $||f'||_1 = 0$. To remedy this, we derive a consequence of our inequality by applying it to $\psi \cdot f$ where ψ is a cut-off function. Since ψ has compact support, we can drop that assumption about f. As long as f is C^∞, $\psi \cdot f \in \mathcal{D}$ and so

$$||\psi f||_\infty \le ||(\psi f)'||_1.$$

Now

$$(\psi f)' = \psi' f + \psi f'$$

and

$$||\psi' f + \psi f'||_1 \le ||\psi' f||_1 + ||\psi f'||_1$$

by the triangle inequality (which is easy to establish for the L^1 norm). We also have the elementary observation

$$\left| \int f(x)g(x)\,dx \right| \le ||g||_\infty \int |f(x)|\,dx$$

which can be written

$$||fg||_1 \le ||f||_1\,||g||_\infty.$$

Thus, altogether, we have

$$||\psi f||_\infty \le ||\psi'||_\infty ||f||_1 + ||\psi||_\infty ||f'||_1.$$

If we create a family of cut-off functions $\psi_t(x) = \psi(tx)$ so that $\psi_t(x) \to 1$ as $t \to 0$, then $||f||_\infty = \lim_{t\to 0} ||\psi_t f||_\infty$ and $||\psi_t||_\infty \le 1$ and $||\psi_t'||_\infty \le t \le 1$, so we have

$$||f||_\infty \le ||f||_1 + ||f'||_1$$

which is exactly the L^1 Sobolev inequality.

While we have succeeded in establishing the desired inequality (at least for smooth functions), we seem to have gotten less than nowhere in proving the more interesting parts of the theorem. We would like to show that if f and f' (in the distributional sense) are integrable, then f is continuous, but we started out assuming that f is differentiable! However, there is a tremendous power concealed in inequalities such as the L^1 Sobolev inequality, which will be illustrated here. Start with f integrable. Regard it as a distribution, and apply the convolution with approximate identity and multiplication by cut-off function argument we gave in section 6.6 to approximate it by test functions. If f and f' are integrable, then the L^1 Sobolev inequality implies that the test functions converge to f uniformly, and uniform limits of continuous functions are continuous. In fact we have already given the cut-off function argument, so let us just consider the convolution argument. Let φ_ϵ denote an approximate identity in $\mathcal{D}$. Then $\varphi_\epsilon * f$ is C^∞. Futhermore $\varphi_\epsilon * f$ is integrable with $||\varphi_\epsilon * f||_1 \le ||f||_1$. To see this just estimate

$$\begin{aligned} ||\varphi_\epsilon * f||_1 &= \int \left| \int \varphi_\epsilon(x-y)f(y)\,dy \right| dx \\ &\le \iint \varphi_\epsilon(x-y)|f(y)|\,dy\,dx \end{aligned}$$

and interchange the order of integration, using $\int \varphi_\epsilon(x-y)\,dx = 1$. What is more, $(\varphi_\epsilon * f)' = \varphi_\epsilon * f'$ is also integrable with $||(\varphi_\epsilon * f)'||_1 \le ||f'||_1$. Thus

$$\begin{aligned} ||\varphi_\epsilon * f||_\infty &\le ||\varphi_\epsilon * f||_1 + ||(\varphi_\epsilon * f)'||_1 \\ &\le ||f||_1 + ||f'||_1 \end{aligned}$$

so in the limit we obtain that f is bounded and satisfies the L^1 Sobolev inequality, and then

$$||\varphi_\epsilon * f - f||_\infty \le ||\varphi_\epsilon * f - f||_1 + ||\varphi_\epsilon * f' - f'||_1.$$

The right side goes to zero as $\epsilon \to 0$ by the approximate identity theorem, and so $\varphi_\epsilon * f \to f$ uniformly, as claimed.

This essentially completes the proof when $n = 1$, except for the observation that if $(d/dx)^j f$ is integrable for $j = 0, 1, \ldots, k+1$ then by applying the previous argument to the functions $(d/dx)^j f$, $j = 0, \ldots, k$ we can show these are continuous, hence $f \in C^k$.

The argument in higher dimensions is similar, except that we have to apply the one-dimensional fundamental theorem of calculus n times. The case $n = 2$ will illustrate the idea without requiring complicated notation. Assume $f \in \mathcal{D}(\mathbb{R}^2)$. We begin with $\partial^2 f/\partial x \partial y$ and integrate in the y-variable to obtain

$$\frac{\partial f}{\partial x}(x,y) = \int_{-\infty}^y \frac{\partial^2 f}{\partial x \partial y}(x,t)\,dt$$

for any fixed x and y. We then integrate in the x-variable to obtain

$$f(x,y) = \int_{-\infty}^x \int_{-\infty}^y \frac{\partial^2 f}{\partial x \partial y}(s,t)\,dt\,ds.$$

We can then estimate

$$
\begin{aligned}
|f(x,y)| &\leq \int_{-\infty}^{x}\int_{-\infty}^{y}\left|\frac{\partial^2 f}{\partial x \partial y}(s,t)\right|\,dt\,ds \\
&\leq \int_{-\infty}^{\infty}\int_{-\infty}^{\infty}\left|\frac{\partial^2 f}{\partial x \partial y}(s,t)\right|\,dt\,ds \\
&= \left\|\frac{\partial^2 f}{\partial x \partial y}\right\|_1
\end{aligned}
$$

and take the supremum over x to obtain

$$
\|f\|_\infty \leq \left\|\frac{\partial^2 f}{\partial x \partial y}\right\|_1 .
$$

This argument required compact support for f, and we remove this hypothesis as before by multiplying by a cut-off function ψ. Now

$$
\frac{\partial^2}{\partial x \partial y}(\psi f) = \psi\frac{\partial^2 f}{\partial x \partial y} + \frac{\partial \psi}{\partial x}\frac{\partial f}{\partial y} + \frac{\partial \psi}{\partial y}\frac{\partial f}{\partial x} + \frac{\partial^2 \psi}{\partial x \partial y}f
$$

so we end up with the inequality

$$
\|f\|_\infty \leq \|f\|_1 + \left\|\frac{\partial f}{\partial x}\right\|_1 + \left\|\frac{\partial f}{\partial y}\right\|_1 + \left\|\frac{\partial^2 f}{\partial x \partial y}\right\|_1 .
$$

This is our L^1 Sobolev inequality for $n = 2$. Notice that it is slightly better than advertised, because it only involves the mixed second derivative $\partial^2 f/\partial x \partial y$, and neither of the pure second derivatives $\partial^2 f/\partial x^2, \partial^2 f/\partial y^2$. The same argument in n dimensions yields the L^1 Sobolev inequality

$$
\|f\|_\infty \leq \sum_{\alpha \in A}\left\|\left(\frac{\partial}{\partial x}\right)^\alpha f\right\|_1
$$

where A is the set of all multi-indexes $\alpha = (\alpha_1, \ldots, \alpha_n)$ where each α_j assumes the value 0 or 1.

The proof of the rest of the theorem in n dimensions is the same as before, except that we have to trade in n derivatives because of the nth-order derivative on the right in the L^1 Sobolev inequality.

The L^1 Sololev inequality is sharp. It is easy to give examples of functions with fewer derivatives in L^1 which are unbounded. Nevertheless, if we have fewer than n derivatives to trade in, we can obtain the same conclusion if these derivatives have finite L^p norm with larger values of p. In fact the rule is that we need more than n/p derivatives. We illustrate this with the case $p = 2$.

Theorem 8.1.2 *(L^2 Sobolev inequality): Suppose*

$$\left\|\left(\frac{\partial}{\partial x}\right)^\alpha f\right\|_2$$

is finite for all α satisfying $|\alpha| \leq N$ for N equal to the smallest integer greater than $n/2$ (so $N = \frac{n+1}{2}$ if n is odd and $N = \frac{n+2}{2}$ if n is even). Then f is continuous and bounded, with

$$\|f\|_\infty \leq c \sum_{|\alpha| \leq N} \left\|\left(\frac{\partial}{\partial x}\right)^\alpha f\right\|_2 .$$

More generally, if

$$\left\|\left(\frac{\partial}{\partial x}\right)^\alpha f\right\|_2$$

is finite for all α satisfying $|\alpha| \leq N + k$, then f is C^k.

We can give a rather quick proof using the Fourier transform. We will show in fact that $\hat{f}$ is integrable, which implies that f is continuous and bounded by the Fourier inversion formula. Since

$$\mathcal{F}\left(\left(\frac{\partial}{\partial x}\right)^\alpha f\right) = (-i\xi)^\alpha \hat{f}(\xi)$$

and

$$\left\|\left(\frac{\partial}{\partial x}\right)^\alpha f\right\|_2 = (2\pi)^{n/2}\left\|\mathcal{F}\left(\left(\frac{\partial}{\partial x}\right)^\alpha f\right)\right\|_2 ,$$

the hypotheses of the theorem imply $\int |\xi^\alpha|^2 |\hat{f}(\xi)|^2 d\xi$ is finite for all $|\alpha| \leq N$. Just taking the cases $\xi^\alpha = 1$ and $\xi^\alpha = \xi_j^N$ and summing we obtain

$$\int \left(1 + \sum_{j=1}^n |\xi_j|^{2N}\right) |\hat{f}(\xi)|^2 \, d\xi < \infty.$$

Now we are going to apply the Cauchy-Schwartz inequality to

$$\hat{f}(\xi) = \left[\hat{f}(\xi)\left(1 + \sum_{j=1}^n |\xi_j|^{2N}\right)^{1/2}\right] \cdot \left[\left(1 + \sum_{j=1}^n |\xi_j|^{2N}\right)^{-1/2}\right] .$$

We obtain

$$\int |\hat{f}(\xi)|\, d\xi \; \le \; \left(\int \left(1 + \sum_{j=1}^{n} |\xi_j|^{2N} \right) |\hat{f}(\xi)|^2\, d\xi \right)^{1/2}$$

$$\cdot \left(\int \left(1 + \sum_{j=1}^{n} |\xi_j|^{2N} \right)^{-1} d\xi \right)^{1/2}.$$

We have already seen that the first integral on the right is finite. We claim that the second integral on the right is also finite, because $2N > n$. The idea is that $\sum_{j=1}^{n} |\xi_j|^{2N} \ge c|\xi|^{2N}$, so

$$\left(1 + \sum_{j=1}^{n} |\xi_j|^{2N} \right)^{-1} \le (1 + c|\xi|^{2N})^{-1}$$

and when we compute the integral in polar coordinates we obtain $c \int_0^{\infty} (1 + r^{2N})^{-1} r^{n-1}\, dr$, which is convergent at infinity when $2N > n$. Thus we have shown that $\hat{f}$ is integrable as claimed, and in fact we have the estimate

$$\int |\hat{f}(\xi)|\, d\xi \le c \sum_{|\alpha| \le N} \left\| \left(\frac{\partial}{\partial x} \right)^{\alpha} f \right\|_2$$

which also proves the L^2 Sobolev inequality since $\|f\|_\infty \le c\|\hat{f}\|_1$.

Notice that we also have the conclusion that f vanishes at infinity by the Riemann-Lebesgue lemma, but this is of relatively minor interest. In fact, the Sobolev inequalities are often used in local form. Suppose we want to show that f is C^k on some bounded open set U. Then it suffices to show

$$\int_V \left| \left(\frac{\partial}{\partial x} \right)^{\alpha} f(x) \right|^2 dx < \infty$$

on all slightly smaller sets V, for $|\alpha| \le N + k$. The reason is that for each $x \in U$ we may apply the L^2 Sobolev inequality to ψf where $\psi \in \mathcal{D}$ is supported on V and $\psi \equiv 1$ on a neighborhood of x. Similarly, if

$$\int_{|x| \le R} \left| \left(\frac{\partial}{\partial x} \right)^{\alpha} f(x) \right|^2 dx < \infty$$

for all $R < \infty$ and all $|\alpha| \le N + k$, then f is C^k on $\mathbb{R}^n$. Of course the same remark applies to the L^1 Sobolev inequality.

The statement of the L^p Sobolev inequality for $p > 1$ is similar to the case $p = 2$, but the proof is more complicated and we will omit it.

Theorem 8.1.3 *(L^p Sobolev inequality):* *Let N_p be the smallest integer greater than n/p, for fixed $p > 1$. If $||(\partial/\partial x)^\alpha f||_p$ is finite for all $|\alpha| \leq N_p$, then f is bounded and continuous and*

$$||f||_\infty \leq c \sum_{|\alpha| \leq N_p} \left\| \left(\frac{\partial}{\partial x}\right)^\alpha f \right\|_p .$$

More generally, if $||(\partial/\partial x)^\alpha f||_p$ is finite for all $|\alpha| \leq N_p + k$, then f is C^k.

The L^p Sobolev inequalities are sharp in the sense that we cannot eliminate the requirement $N_p > n/p$, and we should emphasize that the inequality must be strict. But there is another sense in which they are flabby: if we are required to trade in strictly more than n/p derivatives, what have we gotten in return for the excess $N_p - n/p$? It turns out that we do get something, and we can make a precise statement as long as $\beta = N_p - n/p < 1$. We get Hölder continuity of order β:

$$|f(x) - f(y)| \leq c|x - y|^\beta.$$

The proof of this is not too difficult when $p = 2$ and n is odd, so $N_2 = \frac{n+1}{2}$ and $\beta = \frac{1}{2}$ (if n is even then $\beta = 1$ and the result is false). We will in fact show

$$\frac{|f(x) - f(y)|}{|x - y|^{1/2}} \leq c \sum_{|\alpha| \leq N_2} \left\| \left(\frac{\partial}{\partial x}\right)^\alpha f \right\|_2 .$$

We simply write out the Fourier inversion formula

$$f(x) - f(y) = \frac{1}{(2\pi)^n} \int (e^{-ix\cdot\xi} - e^{-iy\cdot\xi}) \hat{f}(\xi)\, d\xi$$

and estimate

$$
\begin{aligned}
|f(x) - f(y)| &\leq \frac{1}{(2\pi)^n} \int |e^{-ix\cdot\xi} - e^{-iy\cdot\xi}| |\hat{f}(\xi)|\, d\xi \\
&\leq \frac{1}{(2\pi)^n} \left(\int |\hat{f}(\xi)|^2 \left(1 + \sum_{j=1}^n |\xi_j^N| \right)^2 d\xi \right)^{1/2} \\
&\quad \cdot \left(\int |e^{-ix\cdot\xi} - e^{-iy\cdot\xi}|^2 \left(1 + \sum_{j=1}^n |\xi_j^N| \right)^{-2} d\xi \right)^{1/2}
\end{aligned}
$$

using the Cauchy-Schwartz inequality. Since the first integral is finite (dominated by

$$\sum_{|\alpha| \leq N_2} \left\| \left(\frac{\partial}{\partial x}\right)^\alpha f \right\|_2$$

as before), it suffices to show that the second integal is less than a multiple of $|x - y|$. To do this we use

$$\left(1 + \sum_{j=1}^{n} |\xi_j|^N\right)^{-2} \leq c|\xi|^{-n-1}$$

(this is actually a terrible estimate for small values of $|\xi|$, but it turns out not to matter). Then to estimate

$$\int |e^{-ix\cdot\xi} - e^{-iy\cdot\xi}|^2 |\xi|^{-n-1} d\xi$$

we break the integral into two pieces at $|\xi| = |x - y|^{-1}$. If $|\xi| \geq |x - y|^{-1}$ we dominate $|e^{-ix\cdot\xi} - e^{-y\cdot\xi}|^2$ by 4, and obtain

$$\int_{|\xi| \geq |x-y|^{-1}} |e^{-ix\cdot\xi} - e^{-y\cdot\xi}|^2 |\xi|^{-n-1} d\xi$$

$$\leq 4 \int_{|\xi| \geq |x-y|^{-1}} |\xi|^{-n-1} d\xi = c|x - y|$$

after a change of variable $\xi \to |x - y|^{-1}\xi$ (the integral converges because $-n - 1 < -n$). On the other hand, if $|\xi| \leq |x - y|^{-1}$ we use the mean-value theorem to estimate

$$|e^{-ix\cdot\xi} - e^{-iy\cdot\xi}|^2 \leq c|x\cdot\xi - y\cdot\xi|^2 \leq c|x - y|^2|\xi|^2$$

hence

$$\int_{|\xi| \leq |x-y|^{-1}} |e^{-ix\cdot\xi} - e^{-iy\cdot\xi}|^2 |\xi|^{-n-1} d\xi$$

$$\leq c|x - y|^2 \int_{|\xi| \leq |x-y|^{-1}} |\xi|^{-n+1} d\xi = c|x - y|^2$$

after the same change of variable (the integral converges this time because $-n + 1 > -n$). Adding the two estimates we obtain the desired Hölder continuity of order $1/2$.

If you try the same game in even dimensions you will get struck with a divergent integral for $|\xi| \leq |x - y|^{-1}$. What is true in that case is the Zygmund class estimate

$$|f(x + 2y) - 2f(x + y) + f(x)| \leq c|y|$$

which is somewhat weaker than the Lipschitz condition $|f(x + y) - f(x)| \leq c|y|$ which does not hold.

8.2   Sobolev spaces

Because of the importance of the Sobolev inequalities, it is convenient to consider the collection of functions satisfying the hypotheses of such theorems as forming a space of functions, appropriately known as Sobolev spaces. Athough the definition of these spaces has become standard, there seems to be a lack of agreement on how to denote them. Here I will write L_k^p to denote functions with derivatives of order up to k having finite L^p norm. The case $p = 2$ is especially simple, and it is customary to write $L_k^2 = H^k$; I don't know why. The trouble is that H is an overworked letter in mathematical nomenclature, and the symbol H^k could be easily interpreted as meaning something completely different (homology, Hardy space, hyperbolic space, for example, all with a good excuse for the H). The L in our notation stands for *Lebesgue*. Other common notation for Sobolev spaces includes $W^{p,k}$, and the placement of the two indices p and k is subject to numerous changes.

Definition 8.2.1 *The Sobolev space $L_k^p(\mathbb{R}^n)$ is defined to be the space of functions on $\mathbb{R}^n$ such that*

$$\left\|\left(\frac{\partial}{\partial x}\right)^\alpha f\right\|_p$$

is finite for all $|\alpha| \leq k$. Here k is a nonnegative integer and $1 \leq p < \infty$. The Sobolev space norm is defined by

$$\|f\|_{L_k^p} = \left(\sum_{|\alpha|\leq k}\left\|\left(\frac{\partial}{\partial x}\right)^\alpha f\right\|_p^p\right)^{1/p}.$$

When $k = 0$ we write simply L^p.

Of course, when we call it a Sobolev space *norm* we imply that it satisfies the three conditions of a norm described in section 8.1. The Sobolev spaces are examples of *Banach spaces*, which means they are complete with respect to their norms. When $p = 2$, L_k^2 is in fact a *Hilbert space* (this explains the positioning of the pth power in the definition). If you are not familar with these concepts, don't worry; I won't be using them in what follows. On the other hand, they are important concepts, and these are good examples to keep in mind if you decide to study *functional analysis*, as the study of such spaces is called.

The Sobolev inequality theorems can now be stated succinctly as containment relationships between Sobolev spaces and C^k spaces (the space of C^k functions). They say $L_k^p(\mathbb{R}^n) \subseteq C^m(\mathbb{R}^n)$ provided

1. $p = 1$ and $k - n = m$, or

2. $p > 1$ and $k - n/p > m$.

For this reason they are sometimes called the Sobolev *embedding* theorems.

There is another kind of Sobolev embedding theorem which tells you what happens when $k < n/p$, when you do not have enough derivatives to trade in to get continuity. Instead, what you get is a boost in the value of p.

Theorem 8.2.2 $L_k^p(\mathbb{R}^n) \subseteq L_m^q(\mathbb{R}^n)$ *provided* $1 \le p < q < \infty$ *and* $k - n/p \ge m - n/q$. *We have the cooresponding inequality* $\|f\|_{L_m^q} \le c\|f\|_{L_k^p}$.

The theorem is most interesting in the case when we have equality $k - n/p = m - n/q$, in which case it is sharp.

The L^2 Sobolev spaces are easily characterized in terms of Fourier transforms, and this point of view allows us to extend the definition to allow the parameter k to assume noninteger values, and even negative values. We have already used this point of view in the proof of the L^2 Sobolev inequalities. The idea is that since

$$\left(\left(\frac{\partial}{\partial x} \right)^\alpha f \right)^{\hat{}} (\xi) = (-i\xi)^\alpha \hat{f}(\xi)$$

we have

$$\left\| \left(\frac{\partial}{\partial x} \right)^\alpha f \right\|_2 = (2\pi)^{n/2} \|\xi^\alpha \hat{f}(\xi)\|_2$$

by the Plancherel formula; hence

$$\|f\|_{L_k^2} = (2\pi)^{n/2} \left(\int \left(\sum_{|\alpha| \le k} |\xi^\alpha|^2 \right) |\hat{f}(\xi)|^2 \, d\xi \right)^{1/2}.$$

Now the function $\sum_{|\alpha| \le k} |\xi^\alpha|^2$ is a bit complicated, so it is customary to replace it by $(1 + |\xi|^2)^k$. Note that these two functions are of comparable size: there exist constants c_1 and c_2 such that

$$c_1(1 + |\xi|^2)^k \le \sum_{|\alpha| \le k} |\xi^\alpha|^2 \le c_2(1 + |\xi|^2)^k.$$

This means that $f \in L_k^2$ if and only if $\int (1 + |\xi|^2)^k |\hat{f}(\xi)|^2 \, d\xi$ is finite. Also the square root of this integral is comparable in size to $\|f\|_{L_k^2}$. We say it gives an *equivalent* norm.

There is no necessity for k to be an integer in this condition. If $s \geq 0$ is real we may consider those functions in L^2 such that $\int (1 + |\xi|^2)^s |\hat{f}(\xi)|^2 \, d\xi$ is finite to define the Sobolev space L_s^2. Notice that since $1 \leq (1 + |\xi|^2)^s$ the finiteness of this integral implies $\int |\hat{f}(\xi)|^2 \, d\xi < \infty$ which implies $f \in L^2$ by the Plancherel theorem. Thus there is no loss of generality in taking $f \in L^2$ in the first place. If we wish to consider $s < 0$, however, this is not the case. The Sobolev space L_s^2 will then be a space of distributions, not necessarily functions. We say that a tempered distribution f is in L_s^2 for $s < 0$ if $\hat{f}$ corresponds to a function and $\int (1 + |\xi|^2)^s |\hat{f}(\xi)|^2 \, d\xi$ is finite. The spaces L_s^2 decrease in size as s increases. Since $(1 + |\xi|^2)^{s_1} \leq (1 + |\xi|^2)^{s_2}$ if $s_1 \leq s_2$, it follows that $L_{s_2}^2 \subseteq L_{s_1}^2$ (this is true regardless of the sign of s_1 or s_2). Functions in L_s^2 as s increases become increasingly smooth, by the L^2 Sobolev inequalities. Similarly, as s decreases, the distributions in L_s^2 may become increasingly rough.

At least locally, the Sobolev spaces give a scale of smoothness-roughness that describes every distribution. More precisely, if f is a distribution of compact support, there must be some s for which $f \in L_s^2$. We write this as

$$\mathcal{E}' \subseteq \bigcup_s L_s^2.$$

At the other extreme, if $f \in L_s^2$ for every s, then f is C^∞, or

$$\bigcap_s L_s^2 \subseteq \mathcal{E}.$$

The meaning of the Sobolev spaces L_s^2 for noninteger s can be given directly in terms of certain integrated Hölder conditions. For simplicity suppose $0 < s < 1$. A function $f \in L^2$ belongs to L_s^2 if and only if the integral

$$\int_{\mathbb{R}^n} \int_{\mathbb{R}^n} |f(x+y) - f(x)|^2 \frac{dy}{|y|^{n+2s}} \, dx$$

in finite. The idea is that this integral is equal to a multiple of

$$\int |\hat{f}(\xi)|^2 |\xi|^{2s} \, d\xi$$

and it is not hard to see that $\int |\hat{f}(\xi)|^2 (1 + |\xi|^2)^s \, d\xi$ is finite if and only if $\int |\hat{f}(\xi)|^2 d\xi$ and $\int |\hat{f}(\xi)|^2 |\xi|^{2s} d\xi$ are finite.

We begin with the observation

$$\mathcal{F}_x(f(x+y) - f(x))(\xi) = \hat{f}(\xi)(e^{-y \cdot \xi} - 1)$$

and so by the Plancherel formula

$$\int |f(x+y) - f(x)|^2 \, dx = (2\pi)^{-n} \int |\hat{f}(\xi)|^2 |e^{-iy\cdot\xi} - 1|^2 d\xi.$$

Thus, after interchanging the orders of integration, we have

$$\iint |f(x+y) - f(x)|^2 \frac{dy}{|y|^{n+2s}} \, dx$$

$$= (2\pi)^{-n} \int |\hat{f}(\xi)|^2 \left(\int |e^{-y\cdot\xi} - 1|^2 \frac{dy}{|y|^{n+2s}} \right) dx.$$

To complete the argument we have to show

$$\int |e^{-y\cdot\xi} - 1|^2 \frac{dy}{|y|^{n+2s}} = c|\xi|^{2s},$$

which follows from the following three facts:

1. the integral is finite,

2. it is homogeneous of degree $2s$, and

3. it is radial,

because any finite function satisfying 2 and 3 must be a multiple of $|\xi|^{2s}$. The finiteness of the integral follows by analyzing separately the behavior of the integand near 0 and ∞. Near ∞, just use the boundedness of $|e^{-y\cdot\xi} - 1|^2$ and $s > 0$. Near 0, use $|e^{-y\cdot\xi} - 1|^2 \le c|y|^2$ by the mean value theorem, and $s < 1$. The homogeneity follows by the change of variable $y \to |\xi|^{-1}y$ and the fact that the integal is radial (rotation invariant) follows from the observation that the same rotation applied to y and ξ leaves the dot product $y \cdot \xi$ unchanged.

For $s > 1$ we write $s = k + s'$ where k is an integer and $0 < s' < 1$. Then functions in L^2_s are functions in L^2_k whose derivatives of order k are in $L^2_{s'}$, and then we can use the previous characterization.

An interpretation of L^2_s for s negative can be given in terms of duality, in the same sense that $\mathcal{D}'$ is the dual of $\mathcal{D}$. If say $s > 0$, then L^2_{-s} is exactly the space of continuous linear functionals on L^2_s. If $f \in L^2_{-s}$ and $\varphi \in L^2_s$, the value of the linear functional $\langle f, \varphi \rangle$ can be expressed on the Fourier transform side by

$$\langle f, \varphi \rangle = (2\pi)^{-n} \int \hat{f}(\xi) \overline{\hat{f}(\xi)} \, d\xi,$$

at least formally by the Plancherel formula (the appearance of the complex conjugate is slightly different than the convention we adapted for distributions, but it is not really significant). Of course the Plancherel formula is not really valid since f is not in L^2, but we can use the Fourier transform integral as a definition of $\langle f, \varphi \rangle$. The crucial point is that $f \in L^2_{-s}$ and $\varphi \in L^2_s$ exactly implies that the integral exists and is finite. This follows by the Cauchy-Schwartz inequality, if we write

$$\hat{f}(\xi)\overline{\hat{\varphi}(\xi)} = \left[\hat{f}(\xi)(1 + |\xi|^2)^{-s/2}\right]\left[\overline{\hat{\varphi}(\xi)}(1 + |\xi|^2)^{s/2}\right].$$

Then we have

$$\left|\int \hat{f}(\xi)\overline{\hat{\varphi}(\xi)}\, d\xi\right| \leq \left(\int |\hat{f}(\xi)|^2(1 + |\xi|^2)^{-s}\, d\xi\right)^{1/2}$$

$$\left(\int |\hat{\varphi}(\xi)|^2(1 + |\xi|^2)^{s} d\xi\right)^{1/2} \leq c\|f\|_{L^2_{-s}}\|\varphi\|_{L^2_s}.$$

This inequality also proves the continuity of the linear functional.

8.3   Elliptic partial differential equations (constant coefficients)

The Laplacian

$$\Delta = \frac{\partial^2}{\partial x_1^2} + \cdots + \frac{\partial^2}{\partial x_n^2}$$

on $\mathbb{R}^n$ is an example of a partial differential operator that belongs to a class of operators called *elliptic*. A number of remarkable properties of the Laplacian are shared by other elliptic operators. To explain what is going on we restrict attention to constant coefficient operators

$$P = \sum_{|\alpha| \leq m} a_\alpha \left(\frac{\partial}{\partial x}\right)^\alpha$$

where the coefficients a_α are constant (we allow them to be complex). The number m, the highest order of the derivatives involved, is called the *order* of the operator. As we have seen, the Fourier transform of Pu is obtained from $\hat{u}$ by multiplication by a polynomial,

$$(Pu)\,\hat{}\,(\xi) = p(\xi)\hat{u}(\xi)$$

where

$$p(\xi) = \sum_{|\alpha| \leq m} a_\alpha(-i\xi)^\alpha.$$

This polynomial is called the *full symbol* of P, while

$$(p_m)(\xi) = \sum_{|\alpha|=m} a_\alpha(-i\xi)^\alpha$$

is called the *top-order symbol*. (To make matters confusing, the term *symbol* is sometimes used for one or the other of these; I will resist this temptation). For example, the Laplacian has $-|\xi|^2$ as both full and top-order symbol. The operator $I - \Delta$ has full symbol $1 + |\xi|^2$ and top-order symbol $|\xi|^2$. Notice that the top-order symbol is always homogeneous of degree m.

We have already observed that the nonvanishing of the full symbol is extremely useful, for then we can solve the equation $Pu = f$ by taking Fourier transforms and dividing,

$$p(\xi)\hat{u}(\xi) = \hat{f}(\xi) \quad \text{so} \quad \hat{u}(\xi) = \frac{1}{p(\xi)}\hat{f}(\xi).$$

If the full symbol vanishes, this creates problems with the division. However, it turns out that the problems caused by zeroes for small ξ are much more tractable than the problems caused by zeroes for large ξ. We will define the class of elliptic operators so that $p(\xi)$ has zeroes only for small ξ.

Definition 8.3.1 *An operator of order m is called elliptic if the top-order symbol $p_m(\xi)$ has no real zeroes except $\xi = 0$. Equivalently, if the full symbol satisfies $|p(\xi)| \geq c|\xi|^m$ for $|\xi| \geq A$, for some positive constants c and A.*

The equivalence is not immediately apparent, but is not difficult to establish. If $p_m(\xi) \neq 0$ for $\xi \neq 0$, then by homogeneity $|p_m(\xi)| \geq c_1|\xi|^m$ where c_1 is the minimum of $|p(\xi)|$ on the sphere $|\xi| = 1$ (since the sphere is compact, the minimum value is assumed, so $c_1 > 0$). Since $p(\xi) = p_m(\xi) + q(\xi)$ where q is a polynomial of degree $\leq n - 1$, we have $|q(\xi)| \leq c_2|\xi|^{m-1}$ for $|\xi| \geq 1$, so $|p(\xi)| \geq |p_m(\xi)| - |q(\xi)| \geq \frac{1}{2}c_1|\xi|$ if $|\xi| \geq A$ for $A = \frac{1}{2}c_2/c_1$. Conversely, if $|p(\xi)| \geq c|\xi|$ for $|\xi| \geq A$ then a similar argument shows $|p_m(\xi)| \geq \frac{1}{2}c|\xi|$ for $|\xi| \geq A'$, and by homogeneity $p_m(\xi) \neq 0$ for $\xi \neq 0$.

From the first definition, it is clear that ellipticity depends only on the highest order derivatives; for an mth order operator, the terms of order $\leq m-1$ can be modified at will. We will now give examples of elliptic operators in terms of the highest order terms. For $m = 2$, $P = \sum_{|\alpha|=2} a_\alpha (\partial/\partial x)^\alpha$ with a_α real will be elliptic if the quadratic form $\sum_{|\alpha|=2} a_\alpha \xi^\alpha$ is positive (or negative) definite. For $n = 2$, this means the level sets $\sum_{|\alpha|=2} a_\alpha \xi^\alpha =$ constant are ellipses, hence the etymology of the term "elliptic." For $m = 1$, the Cauchy-Riemann operator

$$\frac{\partial}{\partial \bar{z}} = \frac{1}{2}\left(\frac{\partial}{\partial x} + i\frac{\partial}{\partial y}\right)$$

178 8 Sobolev Theory and Microlocal Analysis

in $\mathbb{R}^2 = \mathbb{C}$ is elliptic, because the symbol is $\frac{-i}{2}(\xi + i\eta)$. It is not hard to see that for $m = 1$ there are no examples with real coefficients or for $n \geq 3$. Thus both harmonic and holomorphic functions are solutions to homogeneous elliptic equations $Pu = 0$. For our last example, we claim that when $n = 1$ every operator is elliptic, because $p_m(\xi) = a_m(-i\xi)^m$ vanishes only at $\xi = 0$.

For elliptic operators, we would like to implement the division on the Fourier tansform side algorithm for solving $Pu = f$. We have no problem with division by $p(\xi)$ for large ξ, but we still have possibly small problems if $p(\xi)$ has zeroes for small ξ. Rather than deal with this problem head on, we resort to a strategy that might be cynically described as "defining the problem away." We pose an easier problem that can be easily solved by the technique at hand. Instead of seeking a *fundamental solution*, which is a distribution E satisfying $PE = \delta$ (so $E * f$ solves $Pu = f$, at least for f having compact support, or some other condition that makes the convolution well defined), we seek a *parametrix* (the correct pronunciation of this almost unpronouncable word puts the accent on the first syllable) which is defined to be a distribution F which solves $PF \approx \delta$, for the appropriate meaning of approximate equality. What should this be?

Let us write $PF = \delta + R$. Then the remainder R should be small in some sense, you might think. But this is not quite the right idea. Instead of "small," we want the remainder to be "smooth." The reason is that we are mainly interested in the singularities of solutions, and convolution with a smooth function produces a smooth function, hence it does not contribute at all to the singularities. This point of view is one the key ideas in microlocal analysis and might be described as the *doctrine of microlocal myopia*: pay attention to the singularities, and other issues will take care of themselves.

So we define a distribution F to be a *parametrix* for P if $PF = \delta + R$ where R is a C^∞ function. It is easy to produce a parametrix if P is elliptic. Simply define F via it Fourier transform:

$$\hat{F}(\xi) = \begin{cases} 1/p(\xi) & \text{if } |\xi| \geq A \\ 0 & \text{if } |\xi| < A. \end{cases}$$

Clearly F is a tempered distribution since $|1/p(\xi)| \leq c|\xi|^{-m}$ for $|\xi| \geq A$. Also $(PF)\hat{\ } = \chi(|\xi| \geq A)$ so $\hat{R} = (PF)\hat{\ } - 1 = -\chi(|\xi| \leq A)$ has compact support, hence R is C^∞.

So what good is this parametrix? Well, for one thing, it gives local solutions to the equation $Pu = f$. We cannot just set $u = F * f$, for then $P(F*f) = PF*f = f+R*f$. So instead we try $u = F*g$, where g is closely related to f. Since $P(F * g) = g + R * g$, we need to solve $g + R * g = f$, or rather the local version $g + R * g = f$ on U, for some small open set U.

This problem does not involve the differential operator P anymore, only the integral operator of convolution with R. Here is how we solve it:

To localize to U, we choose a cut-off function $\psi \in \mathcal{D}$ which is one on a neighborhood of U and is supported in a slightly larger open set V. If we can find g such that $g + \psi R * g = \psi f$ on $\mathbb{R}^n$, then on U (where $\psi \equiv 1$) this equation is $g + R * g = f$ as desired. We also need a second, larger cut-off function $\varphi \in \mathcal{D}$ which is one on a neighborhood of V, so that $\varphi \psi = \psi$. If g has support in V, then $\psi g = g$ so we can write our equation as

$$g + \psi R * (\varphi g) = \psi f.$$

Any solution of this equation with support in V will give $u = F * g$ as a solution of $Pu = f$ on U.

Now let $\tilde{R}$ stand for the integral operator $\tilde{R}g = \psi R * (\varphi g)$. Specifically, this is an integral operator

$$\tilde{R}g(x) = \int \psi(x) R(x - y) \varphi(y) g(y) \, dy$$

with kernel $\psi(x) R(x - y)\varphi(y)$, and $\tilde{R}g$ always has support in V since $\psi(x)$ has support in V. We are temped to solve the equation $g + \tilde{R}g = \psi f$ by the perturbation series $g = \sum_{k=0}^{\infty}(-1)^k(\tilde{R})^k(\psi f)$. In fact, if we take the neighborhoods U and V sufficiently small, this series converges to a solution (and the solution is supported in V). The reason for this is that the kernel $\psi(x) R(x-y)\varphi(y)$ actually does become small. The exact condition we need is

$$\iint |\psi(x) R(x - y)\varphi(y)|^2 \, dx \, dy < 1$$

and so the size of the sets U and V depends only on the size of R.

In summary, the parametrix leads to local solutions because the smooth reminder term, when localized, becomes small. Of course this argument did not really use the smoothness of R; it would be enough to have R locally bounded. For a more subtle application of the parametrix we return to the question of location of singularities. If $Pu = f$ and f is singular on a closed set K, what can we say about the singularities of u? More precisely, we say that a distribution f is C^{∞} on an open set U if there exists a C^{∞} function F on U such that $\langle f, \varphi \rangle = \int F(x)\varphi(x) \, dx$ for all test functions supported in U, and we define the *singular support* of f (written sing supp (f)) to be the complement of the union of all open sets on which f is C^{∞}. (Notice the analogy with the definition of "support," which is the complement of the union of all open sets on which f vanishes.) By definition, the singular support is always a closed set. When we ask about the location of the singularities of a distribution, we mean: what is its singular support?

So we can rephrase our question: what is the relationship between sing supp Pu and sing supp u? We claim that there is one obvious containment:

$$\text{sing supp } Pu \subseteq \text{ sing supp } u.$$

The reason for this is that if u is C^∞ on U, then so is Pu, so the complement of sing supp Pu contains the complement of sing supp u; then taking complements reverses the containment. However, it is possible that applying the differential operator P might "erase" some of the singularities of u. For example, if $u(x,y) = h(y)$ for some rough function h, then u is not C^∞ on any open set, so sing supp u is the whole plane. But $(\partial/\partial x)u = 0$ so sing supp $(\partial/\partial x)u$ is empty. This example is made possible because $\partial/\partial x$ is not elliptic in $\mathbb{R}^2$. For elliptic operators we have the identity sing supp $Pu =$ sing supp u. This property is called *hypoellipticity* (a very unimaginative term that means "weaker than elliptic"). It says exactly that every solution to $Pu = f$ will be smooth whenever f is smooth (f is C^∞ on U implies u is C^∞ on U). In particular, if f is C^∞ everywhere then u is C^∞ everywhere. Note that this implies that harmonic and holomorphic functions are C^∞. Even more, it says that if a distribution satisfies the Cauchy-Riemann equations in the distribution sense, then it corresponds to a holomorphic function. This can be thought of as a generalization of the classical theorem that a function that is complex differentiable at every point is continuously differentiable.

To prove hypoellipticity we have to show that if $Pu = f$ is C^∞ on an open set U, then so is u. Now we will immediately localize by multiplying u by $\varphi \in \mathcal{D}$ which is supported in U and identically one on a slightly smaller open set V. Observe that $\varphi u = u$ on V so $P(\varphi u) = g$ with $g = f$ on V. So g is C^∞ on V, and we would like to conclude φu is C^∞ on V, which implies u is C^∞ on V (because $u = \varphi u$ on V). Since V can be varied in this argument, it follows that u is C^∞ on all of U. The point of the localization is that φu and g have compact support.

Reverting to our old notation, it suffices to show that if u and f have compact support and $Pu = f$, then f being C^∞ on an open set U implies u is C^∞ on U. Since u and f have compact support, we can convolve with the parametrix F to obtain $F * Pu = F * f$. Since u has compact support, we have $F * Pu = P(F * u) = PF * u$ (if u did not have compact support $F * u$ might not be defined). Now we can substitute $PF = \delta + R$ to obtain

$$u + R * u = F * f$$

(this is an approximate inverse equation in the reverse order). Since $R * u$ is C^∞ everywhere, the singularities of u and $F * f$ are the same. To complete the proof of hypoellipticity we need to show that convolution with F preserves singularities: if f is C^∞ on U then F is C^∞ on U. In terms of singular supports, this will follow if we can show that sing supp F is the origin, i.e., that F coincides with a C^∞ function away from the origin. Already in this argument we have used the smoothness of the remainder R, and now we need an additional property of the parametrix itself.

Now we have an explicit formula for $\hat{F}$, and this yields a formula for F by the Fourier inversion formula

$$F(x) = \frac{1}{(2\pi)^n} \int_{|\xi| \geq A} e^{-ix \cdot \xi} p(\xi)^{-1} d\xi.$$

The trouble with this formula is that we only know $|p(\xi)^{-1}| \leq c|\xi|^{-m}$, and this in general leads to a divergent integral. Of course this is the way it would have to be because $F(x)$ is usually singular at the origin, and we have not done anything yet to rule out setting $x = 0$. The trick is to multiply by $|x|^{2N}$ for a large value of N (this will tend to mask singularities at the origin). We then have essentially

$$|x|^{2N} F(x) = \frac{1}{(2\pi)^n} \int_{|\xi| \geq A} e^{-ix \cdot \xi} (-\Delta_\xi)^N (p(\xi)^{-1}) \, d\xi$$

(we are ignoring the boundary terms at $|\xi| = A$ which are all C^∞ functions). Now from the fact that P is elliptic we can conclude that

$$|\Delta_\xi^N (p(\xi)^{-1})| \leq c|\xi|^{-m-2N}$$

(for p_m in place of p this follows by homogeneity, each derivative reducing the homogeneity degree by one, and the lower order terms are too small to spoil such an estimate). This decay is fast enough to make the integral converge for $m + 2N > n$, and if $m + 2N > n + k$ it implies that $|x|^{2N} F(x)$ is C^k. This, of course, tells us that $F(x)$ is C^k away from the origin (but not at $x = 0$), and by taking N large we can take k as large as we like. This shows sing supp F is contained in the origin; hence it completes the proof of hypoellipticity.

We can think of hypoellipticity as a kind of qualitative property, since it involves an infinite number of derivatives. There are related quantitative properties, involving finite numbers of derivatives measured in terms of Sobolev spaces. These properties are valid for L^p Sobolev spaces for all p with $1 < p < \infty$, but we will discuss only the case $p = 2$. The idea is that the elliptic operators are honestly of order m in all directions, so there can be no cancellation among derivatives. Since applying an operator of order m loses m derivatives, undoing it should gain m derivatives. This idea is actually false for ordinary derivatives, but it does hold for Sobolev space derivatives.

The properties we are going to describe go under the unlikely name of *a priori estimates* (this Latin mouthful is pronounced ay pree oree).

Theorem 8.3.2 *For P an elliptic operator of order m, if $u \in L^2$ and $Pu \in L^2_k$ then $u \in L^2_{k+m}$ with*

$$\|u\|_{L^2_{k+m}} \leq c(\|u\|_{L^2} + \|Pu\|_{L^2_k}).$$

This is true for any $k \geq 0$ (including noninteger values).

To prove this a priori estimate we work entirely on the Fourier transform side. We are justified in taking Fourier transforms because $u \in L^2$. Now to show $u \in L^2_{k+m}$ we need to show that $\int |\hat{u}(\xi)|^2 (1 + |\xi|^2)^{k+m} \, d\xi$ is finite. We break the integral into two parts, $|\xi| \leq A$ and $|\xi| \geq A$. For $|\xi| \leq A$ we use the fact that $u \in L^2$ and the bound $(1 + |\xi|^2)^{k+m} \leq (1 + A^2)^{k+m}$ to estimate

$$\int_{|\xi| \leq A} |\hat{u}(\xi)|^2 (1 + |\xi|^2)^{k+m} \, d\xi$$

$$\leq \quad (1 + A^2)^{k+m} \int_{|\xi| \leq A} |\hat{u}(\xi)|^2 \, d\xi$$

$$\leq \quad c\|u\|_{L^2}^2.$$

For $|\xi| \geq A$ we use $Pu \in L^2_k$ and $|p(\xi)| \geq c|\xi|^m$ to estimate

$$\int_{|\xi| \geq A} |\hat{u}(\xi)|^2 (1 + |\xi|^2)^{k+m} \, d\xi \quad \leq c \int_{|\xi| \geq A} |p(\xi)\hat{u}(\xi)|^2 (1 + |\xi|^2)^k \, d\xi$$

$$\leq c \int_{\mathbb{R}^n} |Pu\,\hat{}\,(\xi)|^2 (1 + |\xi|^2)^k \, d\xi = c\|Pu\|_{L^2_k}^2.$$

Together, these two estimates prove that $u \in L^2_{k+m}$ and give the desired a priori estimate.

The hypothesis $u \in L^2$ perhaps seems unnatural, but the theorem is clearly false without it. There are plenty of global harmonic functions, $\Delta u = 0$ so $\Delta u \in L^2_k$ for all k, but none of them are even in L^2. You should think of $u \in L^2$ as a kind of minimal smoothness hypothesis (it could be weakened to just $u \in L^2_{-N}$ for some N); once you have this minimal smoothness, the exact Sobolev space L^2_{k+m} for u is given by the exact Sobolev space L^2_k for Pu, and we get the gain of m derivatives as predicted.

The a priori estimates can also be localized, but the story is more complicated. Suppose $u \in L^2, Pu = f$, and f is in L^2_k on an open set U (just restrict all integrals to U). Then u is in L^2_{k+m} on a smaller open set V. The proof of this result is not easy, however, because when we try to localize by multiplying u by φ, we lose control of $P(\varphi u)$. On the set where φ is identically one we have $P(\varphi u) = Pu = f$, but in general the product rule for derivatives produces many terms, $P(\varphi u) = \varphi Pu +$ terms involving lower order derivatives of u. The idea of the proof is that the lower order terms are controlled by Pu also, but we cannot give the details of this argument.

We have seen that the parametrix, and the ideas associated with it, yield a lot of interesting information. However, I will now show that it is possible to construct a fundamental solution after all. We will solve $Pu = f$ for $f \in \mathcal{D}$ as $u = \mathcal{F}^{-1}(p^{-1}\hat{f})$ with the appropriate modifications to take care of the zeroes of p. It can be shown that the solution is of the form $E * f$ for a distribution E (not necessarily tempered) satisfying $PE = \delta$, and then that $u = E * f$ solves $Pu = f$ for any $f \in \mathcal{E}'$. Thus we really are constructing a fundamental solution.

We recall that the Paley-Wiener theorem tells us that $\hat{f}$ is actually an analytic function. We use this observation to shift the integral in the Fourier inversion formula into the complex domain to get around the zeroes of p. We write $\xi' = (\xi_2, \ldots, \xi_n)$ so $\xi = (\xi_1, \xi')$. Then we will write the Fourier inversion formula as

$$\frac{1}{(2\pi)^n} \int_{\mathbb{R}^{n-1}} \left(\int_{-\infty}^{\infty} \frac{\hat{f}(\xi_1, \xi')}{p(\xi_1, \xi')} e^{-ix_1\xi_1} d\xi_1 \right) e^{-ix'\cdot\xi'} d\xi'$$

and we do all the modifications on the inner, one-dimensional integral. We first fix ξ'. Then $p(\zeta, \xi')$ is a polynomial of degree m in the single complex variable ζ. It has at most m distinct complex zeroes, and if γ is a path in the complex plane that does not contain any of these zeroes, and that coincides with the real axis outside the interval $[-A, A]$, such as

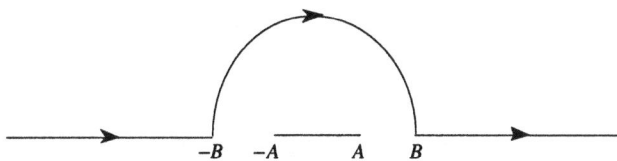

$$-B \quad -A \qquad A \quad B$$

Figure 8.1

then the integral

$$\int_\gamma \frac{\hat{f}(\zeta, \xi')}{p(\zeta, \xi')} e^{-ix_1\zeta} d\zeta$$

converges absolutely. Indeed, for the portion of the integral that coincides with the real axis we may use the estimate $|p(\xi)^{-1}| \leq c|\xi|^{-m}$ and the rapid decay of $\hat{f}(\xi)$ to estimate

$$\int_{|\xi| \geq B} \left| \frac{\hat{f}(\xi_1, \xi')}{p(\xi_1, \xi')} \right| d\xi \leq c(1 + |\xi'|)^{-N}$$

for any N. For the remainder of the contour (the semicircle) we encounter no zeroes of p, so the integrand is bounded and the path is finite.

The particular contour we choose will depend on ξ', so we denote it by $\gamma(\xi')$. The formula for u is thus

$$u(x) = \frac{1}{(2\pi)^n} \int_{\mathbb{R}^{n-1}} \left(\int_{\gamma(\xi')} \frac{\hat{f}(\zeta, \xi')}{p(\zeta_1, \xi')} e^{-ix_1\zeta} d\zeta \right) e^{-ix'\cdot\xi'} d\xi'.$$

When $|\xi'| \geq A$ we choose $\gamma(\xi')$ to be just the real axis, while for $|\xi'| \leq A$ we choose $\gamma(\xi')$ among $m + 1$ contours as described above with $B = A, A + 1, \ldots, A + m$. Since the semicircles are all of distance at least one apart, at least one of them must be at a distance at least $\frac{1}{2}$ from all the m zeroes of $p(\zeta, \xi')$. We choose that contour (or the one with the smallest semicircle if there is a choice). This implies not only that $p(\zeta, \xi')$ is not zero on the contours chosen, but there is a universal bound for $|p(\zeta, \xi')^{-1}|$ over all the semicircular arcs, and there is a universal bound for the lengths of these arcs (they are chosen from a finite number) and for the terms $f(\zeta, \xi')$ and $e^{-ix_1\zeta}$ along these arcs. Thus the integral defining $u(x)$ converges and may be differentiated with respect to the x variables. When we do this differentiation to compute Pu, we produce exactly a factor of $p(\zeta, \xi')$, which cancels the same factor in the denominator:

$$Pu(x) = \frac{1}{(2\pi)^n} \int_{\mathbb{R}^{n-1}} \left(\int_{\gamma(\xi')} \hat{f}(\zeta, \xi') e^{-ix_1\xi} d\zeta \right) e^{-ix'\cdot\xi'} d\xi'.$$

But because $\hat{f}(\zeta, \xi') e^{-ix_1\xi}$ is an analytic function in ζ we use Cauchy's theorem to replace the contour $\gamma(\xi')$ by the real axis

$$\int_{\gamma(\xi')} \hat{f}(\zeta, \xi') e^{-ix_1\zeta} d\zeta = \int_{-\infty}^{\infty} \hat{f}(\xi_1, \xi') e^{-ix_1\xi_1} d\xi_1$$

and hence we have

$$\begin{aligned} Pu(x) &= \frac{1}{(2\pi)^n} \int_{\mathbb{R}^{n-1}} \int_{-\infty}^{\infty} \hat{f}(\xi_1, \xi') e^{-ix_1\xi_1} d\xi_1 e^{-ix'\cdot\xi'} d\xi' \\ &= \frac{1}{(2\pi)^n} \int_{\mathbb{R}^n} \hat{f}(\xi) e^{-ix\cdot\xi} d\xi \end{aligned}$$

which is $f(x)$ by the Fourier inversion formula.

Observe that we were able to shift the contour $\gamma(\xi')$ only after we had applied P to eliminate p in the denominator. For the original integral defining u, the zeroes of p prevent you from shifting contours at will. Because we move into the complex domain, the fundamental solution E constructed may not be given by a tempered distribution. More elaborate arguments show that fundamental solutions can be constructed that are in $\mathcal{S}'$, in fact for any constant coefficient operator (not necessarily elliptic). Although we have constructed a fundamental solution explicitly, the result is too complicated to have any algorithmic significance, except in very special cases.

8.4   Pseudodifferential operators

Our discussion of differential operators, up to this point, has been restricted to constant coefficient operators. For elliptic operators, everything we have done remains true for variable coefficient operators, if properly interpreted. However, the proofs are not as easy, so in this section we will only be able to give the broad outlines of the arguments. As a reward for venturing beyond the comfortable confines of the constant coefficient case, we will get a peek at the theory of pseudodifferential operators, which is one of the glorious achievements of mathematical analysis in the last quarter century.

A variable coefficient linear partial differential operator of order m has the form

$$Pu(x) = \sum_{|\alpha| \leq m} a_\alpha(x) \left(\frac{\partial}{\partial x} \right)^\alpha u(x)$$

with the coefficients $a_\alpha(x)$ being functions. We will assume $a_\alpha(x)$ are C^∞ functions. (There is life outside the C^∞ category, but it is considerably harder.) One very naive way of thinking of such an operator is to "freeze the coefficients." We fix a point $\bar{x}$ and evaluate the functions $a_\alpha(x)$ at $\bar{x}$ to obtain the constant coefficient operator

$$\sum_{|\alpha| \leq m} a_\alpha(\bar{x}) \left(\frac{\partial}{\partial x} \right)^\alpha .$$

As $\bar{x}$ varies we obtain a family of constant coefficient operators, and we are tempted to think that the behavior of the variable coefficient operator should be an amalgam of the behaviors of the constant coefficient operators. This kind of wishful thinking is very misleading; not only does it lead to incorrect conjectures, but it makes us overlook some entirely new phenomena that only show up in the variable coefficient setting. However, for elliptic operators this approach works very well.

We define the full symbol and top-order symbol by

$$p(x, \xi) = \sum_{|\alpha| \leq m} a_\alpha(x)(-i\xi)^\alpha$$

and

$$p_m(x, \xi) = \sum_{|\alpha| = m} a_\alpha(x)(-i\xi)^\alpha.$$

Now these are functions of $2n$ variables, which are polynomials of degree m in the ξ variables and C^∞ functions in both x and ξ variables. The operator is called *elliptic* if $p_m(x, \xi)$ does not vanish if $\xi \neq 0$, and *uniformly elliptic* if there exists a positive constant such that

$$|p_m(x, \xi)| \geq \epsilon |\xi|^m$$

(such an estimate always holds for fixed x, and in fact for x in any compact set).

Suppose we pursue the frozen coefficient paradigm and attempt to construct a parametrix. In the variable coefficient case we cannot expect convolution operators, so we will define a parametrix to be an operator F such that

$$PFu = u + Ru$$

where R is an integral operator with C^∞ kernel $R(x, y)$:

$$Ru(x) = \int R(x, y)u(y) \, dy.$$

Since we are in a noncommutative setting we will also want

$$FPu = u + R_1 u$$

where R_1 is an operator of the same type as R. We could try to take for F the parametrix for the frozen coefficient operator at each point:

$$Fu(x) = \frac{1}{(2\pi)^n} \int_{|\xi| \geq A(x)} \frac{\hat{u}(\xi)}{p(x, \xi)} e^{-ix \cdot \xi} d\xi$$

where $A(x)$ is chosen large enough that $p(x, \xi)$ has no zeroes in $|\xi| \geq A(x)$. It turns out that this guess is not too far off the mark.

The formula we have guessed for a parametrix is essentially an example of what is called a *pseudodifferential operator* (abbreviated ψDO). By definition, a ψDO is an operator of the form

$$\frac{1}{(2\pi)^n} \int \sigma(x, \xi)\hat{u}(\xi)e^{-ix \cdot \xi} \, d\xi$$

where $\sigma(x,\xi)$, the *symbol*, belongs to a suitable symbol class. There are actually many different symbol classes in use; we will describe one that is known as the *classical symbols*. (I will not attempt to justify the use of the term "classical" in mathematics; in current usage it seems to mean anything more than five years old.) A classical symbol of order r (any real number) is a C^∞ function $\sigma(x,\xi)$ that has an asymptotic expansion

$$\sigma(x,\xi) \sim \sigma_r(x,\xi) + \sigma_{r-1}(x,\xi) + \sigma_{r-2}(x,\xi) + \cdots$$

where each $\sigma_{r-k}(x,\xi)$ is a homogeneous function of degree $r - k$ in the ξ variables away from the origin. Except for polynomials, it is impossible for a function to be both C^∞ and homogeneous near the origin; thus we require $\sigma_{r-k}(x,t\xi) = t^{r-k}\sigma_{r-k}(x,\xi)$ only for $|\xi| \geq 1$ and $t \geq 1$. The meaning of the asymptotic expansion is that if we take a finite number of terms, they approximate $\sigma(x,\xi)$ in the following sense:

$$\left|\sigma(x,\xi) - \sum_{k=0}^{m} \sigma_{r-k}(x,\xi)\right| \leq c|\xi|^{r-m-1} \text{ for } |\xi| \geq 1.$$

In other words, the difference decays at infinity in ξ as fast as the next order in the asymptotic expansion. (There is also a related estimate for derivatives, with the rate of decay increasing with each ξ derivative, but unchanged by x derivatives.)

The simplest example of a classical symbol is a polynomial in ξ of degree r (r a postive integer), in which case $\sigma_{r-k}(x,\xi)$ is just the homogeneous terms of degree $r - k$. The asymptotic sum is just a finite sum in this case, and we have equality

$$\sigma(x,\xi) = \sigma_r(x,\xi) + \sigma_{r-1}(x,\xi) + \cdots + \sigma_0(x,\xi).$$

The associated ψDO is just the differential operator with full symbol σ: if $\sigma(x,\xi) = \sum_{|\alpha|\leq r} a_\alpha(x)(-i\xi)^\alpha$ then

$$\frac{1}{(2\pi)^n} \int \sigma(x,\xi)\hat{u}(\xi)e^{-ix\cdot\xi}\,d\xi = \sum_{|\alpha|\leq r} a_\alpha(x)\left(\frac{\partial}{\partial x}\right)^\alpha u(x).$$

Thus the class of pseudodifferential operators is a natural generalization of the differential operators.

The theory of pseudodifferential operators embraces the doctrine of microlocal myopia in the following way: two operators are considered equivalent if they differ by an integral operator with C^∞ kernel $\left(\int R(x,y)\,u(y)\,dy\right.$ for R a C^∞ function). Such an operator produces a C^∞ output regardless of the input (u may be a distribution) and is considered trivial. The portion of the symbol corresponding to small values of ξ produces such a trivial operator. For this reason, in describing a ψDO one frequently just specifies the symbol for $|\xi| \geq 1$. (In what follows we will ignore behavior for $|\xi| \leq 1$, so when we say a function is "homogeneous," this means just for $|\xi| \geq 1$.) Also, from this point of view, a parametrix is a two-sided inverse: PF and FP are equivalent to the identity operator.

We now describe, without proof, the basic properties of ψDO's and their symbols:

1. (*Closure under composition*) If P and Q are ψDO's of order r and s then PQ is a ψDO of order $r + s$. (Some technical assumptions are needed to ensure that the composition exists.) If $p(x, \xi)$ and $q(x, \xi)$ are the symbols of P and Q, then the symbol of PQ has the asymptotic expansion

$$\sum_\alpha \frac{1}{\alpha!} \left(\frac{\partial}{\partial x}\right)^\alpha p(x, \xi) \left(-i \frac{\partial}{\partial \xi}\right)^\alpha q(x, \xi).$$

Note that $(\partial / \partial x)^\alpha p(x, \xi)$ is a symbol of order r, and

$$\left(-i \frac{\partial}{\partial \xi}\right)^\alpha q(x, \xi)$$

is a symbol of order $s - |\alpha|$, and their product has order $r + s - |\alpha|$. In particular, the top-order parts of the symbol multiply: $p_r(x, \xi) q_s(x, \xi)$ is the $r + s$ order term in the symbol of PQ. Since multiplication is commutative, the commutator $[P, Q] = PQ - QP$ has order $r + s - 1$. This is a familiar fact for differential operators.

2. (*Symbolic completeness*) Given any formal sum of symbols $\sum_{k=0}^{\infty} p_{r-k}(x, \xi)$ where $p_{r-k}(x, \xi)$ is homogeneous of order $r - k$, there exists a symbol $p(x, \xi)$ of order r that has this asymptotic expansion. This property explains the meaning of the symbolic product for compositions.

3. (*Pseudolocal property*) If u is C^∞ in an open set, then so is Pu. Thus sing supp $Pu \subseteq$ sing supp u. This property is obvious for differential operators and was one of the key properties of the parametrix constructed in section 8.3. Another way of saying this is that a ψDO is represented as an integral operator with a kernel that is C^∞ away from the diagonal. At least formally,

$$Pu(x) = \frac{1}{(2\pi)^n} \int \sigma(x, \xi) \left(\int e^{iy \cdot \xi} u(y) dy\right) e^{-ix \cdot \xi} d\xi$$

$$= \int \left(\frac{1}{(2\pi)^n} \int \sigma(x, \xi) e^{-i(x-y) \cdot \xi} d\xi\right) u(y) \, dy$$

is an integral operator with kernel

$$\frac{1}{(2\pi)^n} \int \sigma(x,\xi)e^{-i(x-y)\cdot\xi}d\xi.$$

The interchange of the orders of integration is not valid in general (it is valid if the order of the operator $r < -n$). However, away from the diagonal, $x - y \neq 0$ and so the exponential $e^{-i(x-y)\cdot\xi}$ provides enough cancellation that the kernel actually exists and is C^∞, if the integral is suitably interpreted (as the Fourier transform of a tempered distribution, for example). As x approaches y, the kernel may blow up, although it can be interpreted as a distribution on $\mathbb{R}^{2n}$.

4. (*Invariance under change of variable*) If $h : \mathbb{R}^n \to \mathbb{R}^n$ is C^∞ with a C^∞ inverse h^{-1}, then $(P(u \circ h^{-1})) \circ h$ is also a ψDO of the same order whose symbol is $\sigma(h(x),(h'(x)^{tr})^{-1}\xi)$ + lower order terms. This shows that the class of pseudodifferential operators does not depend on the coordinate system, and it leads to a theory of ψDO's on manifolds where the top-order symbol is interpreted as a section of the cotangent bundle.

5. (*Closure under adjoint*) The adjoint of a ψDO is a ψDO of the same order whose symbol is $\overline{\sigma(x,\xi)}$+ lower order terms.

6. (*Sobolev space estimates*) A ψDO of order r takes L^2 Sobolev spaces L^2_s into L^2_{s-r}, at least locally. If $u \in L^2_s$ and u has compact support, then $\psi Pu \in L^2_{s-r}$ if $\psi \in \mathcal{D}$. With additional hypotheses on the behavior of the symbol in the x-variable, we have the global estimate

$$||Pu||_{L^2_{s-r}} \leq c||u||_{L^2_s}.$$

Using these properties of ψDO's, we can now easily construct a parametrix for an elliptic partial differential operator of order m, which is a ψDO of order $-m$. If

$$p(x,\xi) = p_m(x,\xi) + p_{m-1}(x,\xi) + \cdots + p_0(x,\xi)$$

is the symbol of the given differential operator P, write

$$q(x,\xi) \sim q_{-m}(x,\xi) + q_{-m-1}(x,\xi) + \cdots$$

for the symbol of the desired parametrix Q. We want the composition QP to be equivalent to the identity operator, so its symbol should be 1. But the symbol of QP is given by the asymptotic formula

$$\sum_\alpha \frac{1}{\alpha!} \left(\frac{\partial}{\partial x}\right)^\alpha q(x,\xi) \left(-i\frac{\partial}{\partial\xi}\right)^\alpha p(x,\xi).$$

Note that this is actually a finite sum, since

$$\left(-i\frac{\partial}{\partial \xi}\right)^{\alpha} p(x,\xi) = 0$$

if $|\alpha| > m$ because $p(x,\xi)$ is a polynomial of degree m in ξ. Substituting the asymptotic expansion for q and the sum for p yields the asymptotic formula

$$\sum_{|\alpha| \leq m} \sum_{k=0}^{\infty} \sum_{j=0}^{m} \frac{1}{\alpha!} \left(\frac{\partial}{\partial x}\right)^{\alpha} q_{-m-k}(x,\xi) \left(-i\frac{\partial}{\partial \xi}\right)^{\alpha} p_{m-j}(x,\xi)$$

where each term is homogeneous of degree $-k-j-|\alpha|$. Thus there are only a finite number of terms of any fixed degree of homogeneity, the largest of which being zero, for which there is just the one term $q_{-m}(x,\xi)P_m(x,\xi)$. Since we want the symbol to be 1, which has homogeneity zero, we set $q_{-m}(x,\xi) = 1/p_m(x,\xi)$, which is well defined because P is elliptic, and has the correct homogeneity $-m$. All the other terms involving q_{-m} have lower order homogeneity. We next choose q_{-m-1} to kill the sum of the terms of homogeneity -1. These terms are just

$$q_{-m-1}p_m + q_{-m}p_{m-1} + \sum_{|\alpha|=1} \left(\frac{\partial}{\partial x}\right)^{\alpha} q_{-m} \left(-i\frac{\partial}{\partial \xi}\right)^{\alpha} p_m,$$

and we can set this equal to zero and solve for q_{-m-1} (with the correct homogeneity) again because we can divide by p_m. Continuing this process inductively, we can solve for q_{-m-k} by setting the sum of terms of order $-k$ equal to zero, since this sum contains $q_{-m-k}p_m$ plus terms already determined.

This process gives us an asymptotic expansion of $q(x,\xi)$. By the symbolic completeness of ψDO symbols, this implies that there actually is a symbol with this expansion and a corresponding $\psi DO\, Q$. By our construction QP has symbol 1, hence it is equivalent to the identity as desired. A similar computation shows that PQ has symbol 1, so Q is our desired parametrix.

Once we have the existence of the parametrix, we may deduce the equivalents of the properties of constant coefficient elliptic operators given in section 8.3 by essentially the same reasoning. The local existence follows exactly as before, because once we localize to a sufficiently small neighbohood, the remainder term becomes small. The hypoellipticity follows from the pseudo-local property of the parametrix. Local a priori estimates follow from the Sobolev space estimates for ψDO's (global a priori estimates require additional global assumptions on the coefficients). However, there is no analog of the fundamental solution. Elliptic equations are subject to the "index" phenomenon: existence and uniqueness in expected situations may fail by finite-dimensional spaces. We will illustrate this shortly.

Most situations in which elliptic equations arise involve either closed manifolds (such as spheres, tori, etc.) or bounded regions with boundary. If Ω is a bounded open set in $\mathbb{R}^n$ with a regular boundary Γ, a typical problem would be to solve $Pu = f$ on Ω with certain boundary conditions, the values of u and some of its derivatives on the boundary, as given. Usually, the number of boundary conditions is half the order of P, but clearly this expectation runs into difficulties for the Cauchy-Riemann operator. We will not be able to discuss such boundary value problems here, except to point out that one way to approach them is to reduce the problem to solving certain pseudodifferential equations on the boundary.

One example of a boundary value problem in which the index phenomenon shows up is the Neumann problem for the Laplacian. Here we seek solutions of $\Delta u = 0$ on Ω with $\partial u/\partial n = g$ given on Γ ($\partial/\partial n$ refers to the normal derivative, usually in the outward direction). To be specific, suppose Ω is the disk $|x| < 1$ in $\mathbb{R}^2$ and Γ the unit circle, so $\partial/\partial n$ is just the radial derivative in polar coordinates. It follows from Green's theorem that $\int_0^{2\pi} g(r, \theta)\, d\theta = 0$ is a necessary and sufficient condition for the solution to exist, and the solution is not unique since we can always add a constant. Here we have a one-dimensional obstruction to both existence and uniqueness, and the index is zero (it is defined to be the difference of the two dimensions). The reason the index is considered, rather than the two dimensions separately, is that it is relatively robust. It is unchanged by lower order terms, and in fact depends only on certain topological properties of the top-order symbol. The famous Atiyah-Singer Index Theorem explains the relationship between the analytically defined index and the topological quantities in the general situation.

8.5   Hyperbolic operators

Another important class of partial differential operators is the class of hyperbolic operators. Before we can describe this class we need to introduce the notion of *characteristics*. Suppose $P = \sum_{|\alpha| \le m} a_\alpha(x) \left(\partial/\partial x\right)^\alpha$ is an operator of order m. We may ask in which directions it is really of order m. For ordinary differential operators $a_m(x) \left(d/dx\right)^m + \cdots + a_0(x)$, the obvious condition $a_m(x) \ne 0$ is all that we need to be assured that the operator is everywhere of order m. But in higher dimensions, the answer is more subtle. An operator like the Laplacian $\left(\partial^2/\partial x^2\right) + \left(\partial^2/\partial y^2\right)$ in $\mathbb{R}^2$ is clearly of order 2 in both the x and y directions since it involves pure second derivatives in both variables, while the operator $\partial^2/\partial x \partial y$ is not of order 2 in either of these directions. Still, it is a second-order operator, and if we take any slanty direction it will be of second order. For example, the change of variable $x + y = t, x - y = s$, transforms $\partial/\partial x$ into

$$\frac{\partial t}{\partial x}\frac{\partial}{\partial t} + \frac{\partial s}{\partial x}\frac{\partial}{\partial s} = \frac{\partial}{\partial t} + \frac{\partial}{\partial s}$$

and $\partial/\partial y$ into

$$\frac{\partial t}{\partial y}\frac{\partial}{\partial t} + \frac{\partial s}{\partial y}\frac{\partial}{\partial s} = \frac{\partial}{\partial t} - \frac{\partial}{\partial s},$$

hence $\partial^2/\partial x \partial y$ into $(\partial^2/\partial t^2) - (\partial^2/\partial s^2)$ which is clearly of order 2 in both variables.

In general, given an operator P, a point x, and a direction v, we make an orthogonal change of variable so that v lies along one of the axes, say the x_1 axis. If the coefficient of $(\partial/\partial x_1)^m$ (in the new coordinte system) is nonzero at the point x, then it is reasonable to say that P is of order m at the point x in the direction v. We use the term *nonchacracteristic* to describe this situation, so *characteristic* means the coefficient of $(\partial/\partial x_1)^m$ is zero at x. From this point of view, the notation $\partial/\partial x_j$ for partial derivatives is unfortunate. It is incomplete in the sense that $\partial/\partial x_1$ means the rate of change as x_1 is varied while $x_2, \ldots, x_n$ are held fixed. It is much better to think in terms of directional derivatives, say

$$d_v f(x) = \lim_{h \to 0} \frac{1}{h}(f(x + hv) - f(x)).$$

If $v_1, \ldots, v_n$ is any linearly independent basis of $\mathbb{R}^n$ (not necessarily orthogonal), then the chain rule implies that any first-order partial derivative is a linear combination of $d_{v_1}, \ldots, d_{v_n}$. Thus any linear partial differential operator of order m can be written as $\sum_{|\alpha| \le m} b_\alpha(x)(d_v)^\alpha$ where

$$(d_v)^\alpha = (d_{v_1})^{\alpha_1}(d_{v_2})^{\alpha_2} \ldots (d_{v_n})^{\alpha_n}.$$

At a given point x, the characteristic directions v are those for which $b_\alpha(x) = 0$ if $\alpha = (m, 0, \ldots, 0)$ and $v_1 = v$, while the noncharacteristic directions are those for which $b_\alpha(x) \ne 0$. It is easy to see by the chain rule that this definition only depends on the direction $v = v_1$ and not on the other directions $v_2, \ldots, v_n$ in the basis.

Still, it is rather awkward to have to recompute the operator in terms of a new basis for every direction v in order to settle the issue. Fortunately, there is a much simpler approach. We claim that *the characteristic directions are exactly those for which the top-order symbol vanishes*, $p_m(x, v) = 0$. This is obvious if $v = (1, 0, \ldots, 0)$, because $p_m(x, v) =$

$\sum_{|\alpha|=m} a_\alpha(x)(-iv)^\alpha$ and for this choice of v we have $(-iv)^\alpha = 0$ if any factor $v_2, \ldots, v_n$ appears to a nonzero power. Thus the single term corresponding to $\alpha = (m, 0, \ldots, 0)$ survives (lower order terms are not part of the top-order symbol), $p_m(x, v) = (-i)^m a_\alpha(x)$ and we are back to the previous definition. More generally, if

$$\sum_{|\alpha|=m} a_\alpha(x) \left(\frac{\partial}{\partial x}\right)^\alpha = \sum_{|\beta|=m} b_\beta(x)(d_v)^\beta$$

for an orthonormal basis $v_1, \ldots, v_n$ with $v_1 = v$ then

$$\sum_{|\alpha|=m} a_\alpha(x)(-i\xi)^\alpha = \sum_{|\beta|=m} b_\beta(x)(-iv \cdot \xi)^\alpha$$

and so $p(x, v) = (-i)^m b_\beta(x)$ for $\beta = (m, 0, \ldots, 0)$ so $b_\beta(x) \neq 0$ if and only if $p(x, v) \neq 0$.

With this criterion it is easy to compute characteristics. For example, the definition of an elliptic operator is exactly that there are no characteristic directions. For the operator $\partial^2/\partial x \partial y$ the x and y axes are the only characteristic directions. For the wave operator $(\partial^2/\partial t^2) - k^2 \Delta x$, the characteristic directions are of the form (τ, ξ) where $\tau = \pm k|\xi|$, and these form a double cone. For constant coefficient operators the characteristic directions do not depend on the point x, but this is not true in general.

Why is it important to know which directions are characteristic? For one thing, it helps us understand the nature of boundary conditions. Suppose S is a smooth hypersurface in $\mathbb{R}^n$ (hypersurface means of dimension $n - 1$, hence codimension 1). The simplest example is the flat hypersurface $x_n = 0$. In fact, the implicit function theorem says that locally every smooth hypersurface can be given by the equation $x_n = 0$ in the appropriate coordinate system (of course for this we must allow curvilinear coordinate systems). At each point of the surface we have a normal direction (we have to choose among two opposites) and $n - 1$ tangential directions. For the surface $x_n = 0$ the normal direction is $(0, \ldots, 0, 1)$. We say that the hypersurface S is *noncharacteristic* for P if at each point x in S, the normal direction is noncharacteristic. This means exactly that the differential operator is of order m in the normal direction.

For noncharacteristic surfaces, it makes sense to consider the *Cauchy problem*: find solutions to $Pu = f$ that satisfy

$$u\Big|_S = g_0, \quad \frac{\partial u}{\partial n}\Big|_S = g_1, \ldots \left(\frac{\partial}{\partial n}\right)^{m-1} u\Big|_S = g_{m-1}$$

where f (defined on $\mathbb{R}^n$) and $g_0, g_1, \ldots, g_m$ (defined on S) are called the *Cauchy data*. The rationale for giving this sort of data is the following.

Once we know the value of u on S, we can compute all tangential first derivatives, so the only first derivative it makes sense to specify is the normal one. Once all first derivatives are known on S, we can again differentiate in tangential directions. Thus, among second derivatives, only $(\partial^2/\partial n^2)u$ remains to be specified on S. If we continue in this manner, we see that there are no obvious relationships among the Cauchy data, and together they determine all derivatives or order $\leq m - 1$ on S. Now we bring in the differential equation. Because S is noncharacteristic, we can solve for $(\partial/\partial n)^m u$ on S in terms of derivatives already known on S. By repeatedly differentiating the differential equation, we eventually obtain the value of all derivatives of u on S, with no consistency conditions arising from two different computations of the same derivative. In other words, the Cauchy data gives exactly the amount of information, with no redundancy, needed to determine u to infinite order on S.

There is still the issue of whether knowing u to infinite order on S allows us to solve the differential equation off S. In the real analytic case this is the content of the famous Cauchy-Kovalevska Theorem. We have to assume that all the Cauchy data, and the coefficients of P as well, are real analytic functions (that means they have convergent power series expansions locally). The theorem says that then there exist solutions of the Cauchy problem locally (in a small enough neighborhood of any point of S), the solutions being real analytic and unique. This is an example of a theorem that is too powerful for its own good. Because it applies to such a large class of operators, its conclusions are too weak to be very useful. In general, the solutions may exist in only a very small neighborhood of S, it may not depend continuously on the data, and it may fail to exist altogether if the data is not analytic (even for C^∞ data).

For this reason, it seems worthwhile to try to find a smaller class of operators for which the Cauchy problem is well-posed. (By *well-posed* we mean that a solution exists, is unique, and depends continuously on the data.)

This is the structural motivation for the class of hyperbolic equations. Of course, to be useful, the definition of "hyperbolic" must be formalistic, only involving properties of the symbol that are easily checked. In order to simplify the discussion, we begin by assuming that the operator has constant coefficients and contains only terms of the highest order m. In other words, the full symbol is a polynomial $p(\xi)$ that is homogeneous of degree m, hence equal to its top-order symbol.

To give the definition of hyperbolic in this context, we look at the polynomial in one complex variable $z, p(\xi + zv)$ for each fixed ξ and v a given direction. This is a polynomial of degree $\leq m$, and is exactly of degree m when v is noncharacteristic, the coefficient of z^m being $p(v)$. In that case we can factor the polynomial

$$p(\xi + zv) = p(v) \prod_{j=1}^{m} (z - z_j(\xi))$$

where $z_j(\xi)$ are the complex roots. We say P is *hyperbolic* in the direction v, if v is a noncharacteristic direction and all the roots $z_j(\xi)$ are real, for all ξ.

The prototype example is the wave equation, which is hyperbolic in the t-direction. The notation is slightly different, in that $p(\tau, \xi) = -\tau^2 + k^2 |\xi|^2$, so $v = (1, 0, \ldots, 0)$ is the t-direction. Then

$$p((\tau, \xi) + zv) = p(\tau + z, \xi) = -(\tau + z)^2 + k^2 |\xi|^2 = -(z + \tau + k|\xi|)(z + \tau - k|\xi|)$$

which clearly shows the two real zeroes.

For hyperbolic operators of the type we are considering, it is easy to write down a fundamental solution:

$$u(x) = \frac{1}{(2\pi)^n} \int \frac{\hat{f}(\xi + i\lambda v)}{p(\xi + i\lambda v)} e^{-ix \cdot (\xi + i\lambda v)} \, d\xi$$

for $\lambda \neq 0$ real is well defined and solves $Pu = f$ for any $f \in \mathcal{D}$. The idea is that we do not encounter any zeroes of the symbol, and in fact we can estimate $p(\xi + i\lambda v)$ from below; since $p(\xi + i\lambda v) = p(v) \prod_{j=1}^{m} (i\lambda - z_j(\xi))$ and $z_j(\xi)$ is real so $|i\lambda - z_j(\xi)| \geq |\lambda|$ we obtain $|p(\xi + i\lambda v)| \geq |p(v)||\lambda|^m$. This estimate and the Paley-Wiener estimates for $\hat{f}(\xi + i\lambda v)$ guarantee that the integral defining u converges and that we can differentiate with respect to x any number of times inside the integral. Thus

$$Pu(x) = \frac{1}{(2\pi)^n} \int \hat{f}(\xi + i\lambda v) e^{-ix \cdot (\xi + i\lambda v)} \, d\xi.$$

By using the Cauchy integral formula we can shift the contour (one dimension at a time) back to $\mathbb{R}^n$, and then $Pu = f$ by the Fourier inversion formula. In fact, the same Cauchy integral formula argument can be applied to the integral defining u to shift the value of λ, provided we do not try to cross the $\lambda = 0$ divide (where there are zeroes of the symbol). Thus our construction really only produces two different fundamental solutions, corresponding to $\lambda > 0$ and $\lambda < 0$. We denote them by E^+ and E^-.

These turn out to be very special fundamental solutions. First, by a variant of the Paley-Wiener theorem, we can show that E^+ is supported in the half-space $x \cdot v \geq 0$, and E^- is supported in the other half-space $x \cdot v \leq 0$. (By "support" of $E^\pm$ we mean the support of the distributions that give $E^{\pm}$ by convolution.) But we can say even more. There is an open cone Γ that contains the direction v, with the property that P is hyperbolic with respect to every direction in Γ. The fundamental solutions $E^{\pm}$ are the same for every $v \in \Gamma$, so in fact the support of E^+ is contained in the *dual cone* Γ^* defined by $\{x : x \cdot v \geq 0 \text{ for all } v \in \Gamma\}$, and the support of E^- is contained in $-\Gamma^*$. I use the word *cone* here to mean a subset of $\mathbb{R}^n$ that contains an entire ray $\lambda x(\lambda > 0)$ for every point x in the cone. It can also be shown that the cones Γ and Γ^* are convex. The cone Γ^* is referred to as the *forward light cone*, (or sometimes this term is reserved for the boundary of Γ^*, and Γ^* is called the *interior of the forward light cone*).

There are two ways to describe the cone Γ. The first is to take $\mathbb{R}^n$ and remove all the characteristic directions; what is left breaks up into a number of connected components. Γ is the component containing v. For the other description, we again look at the zeroes of the polynomial $p(\xi + zv)$, which are all real. Then ξ is in Γ if all the real roots are negative (note that v is in Γ according to this definition, since $p(v + zv) = p(v)(z + 1)^m$ which has $z = -1$ as the only root). It is not obvious that these two descriptions coincide nor that P is hyperbolic with respect to all directions in Γ, but these facts can be proved using the algebra of polynomials in $\mathbb{R}^n$.

The fact that Γ is an open cone implies that the dual cone Γ^* is *proper*, meaning that it is properly contained in a half-space. In particular, this means that if we slice the light cone by intersecting it with any hyperplane perpendicular to v, we get a bounded set. This observation will have an interesting interpretation concerning solutions of the hyperbolic equation.

We can illustrate these ideas with the example of the wave equation. The characteristic directions were given by the equation $\tau^2 - k^2|\xi|^2 = 0$ in $\mathbb{R}^{n+1}(\tau \in \mathbb{R}^1, \xi \in \mathbb{R}^n)$. The complement breaks up into 3 regions (or 4 regions if $n = 1$). Γ is the region where $\tau > 0$ and $k|\xi| < |\tau|$. The other two regions are $-\Gamma(\tau < 0$ and $k|\xi| < |\tau|)$ and the outer region where $k|\xi| > |\tau|$.

Notice how the second description of Γ works in this example. The two roots were computed to be $-\tau - k|\xi|$ and $-\tau + k|\xi|$. If these are both to be negative, then by summing we see $\tau > 0$. Then $-\tau - k|\xi| < 0$ is automatic and $-\tau + k|\xi| < 0$ is equivalent to $k|\xi| < |\tau|$.

We can also compute Γ^* for this example. For (t, x) to be in Γ^* we must have $t\tau + x \cdot \xi \geq 0$ for all (τ, ξ) in Γ. Taking $(\tau, \xi) = (1, 0)$ shows that we must have $t \geq 0$. Now since $|x \cdot \xi| \leq |x||\xi| \leq |x|\tau/k$ for (τ, ξ) in Γ we have

$$t\tau + x \cdot \xi \geq t\tau - |x|\tau/k = \tau(t - |x|/k)$$

will be positive if $|x| \leq kt$. However, if $|x| > kt$ we can choose ξ in the direction opposite x to make the inequality an equality, and then we get a negative value. Thus Γ^* is given exactly by the conditions $t \geq 0$ and $|x| \leq kt$.

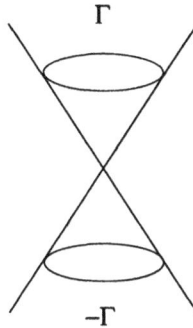

Figure 8.2

Using the fundamental solutions $E^{\pm}$, we can solve the Cauchy problem for any hypersurface S that is *spacelike*. The definition of spacelike is that the normal direction must lie in Γ. The simplest example is the flat hyperplane $x \cdot v = 0$ whose normal direction is v. In the example of the wave equation, if the surface S is defined by the equation $t = h(x)$, then the normal direction is $(1, -\Delta_x h(x))$, so the condition that S be spacelike is that $|\nabla h(x)| \leq 1/k$ everywhere. We will assume that the spacelike surface is complete (it cannot be extended further) and smooth, and that it divides the remaining points of $\mathbb{R}^n$ into two pieces, a "past" and a "future," with the cone Γ pointing in the future direction. In this situation it can be shown that the Cauchy problem is well-posed. Because the argument is intricate we only give a special case.

Suppose we want to solve $Pu = f$,

$$u\Big|_S = 0, \ \frac{\partial}{\partial n} u\Big|_S = 0 \ldots \left(\frac{\partial}{\partial n}\right)^{m-1} u\Big|_S = 0$$

where f is a C^{∞} function with support in the future. We claim that the solution is simply $u = E^+ * f$. We have seen that u solves $Pu = f$ if f has compact support. However, at any fixed point x, $E^+ * f$ only involves the values of f in the set $x - \Gamma^*$. For x in S or in the past, $x - \Gamma^*$ lies entirely in the past where f is zero, so u vanishes on S and the past, hence vanishes to infinite order on S. But for x in the future, $x - \Gamma^*$ intersects the future only in a bounded set, as shown in Figure 8.3.

The convolution is well defined even if f does not have compact support. In fact, this argument shows that the solution u at x in the future depends

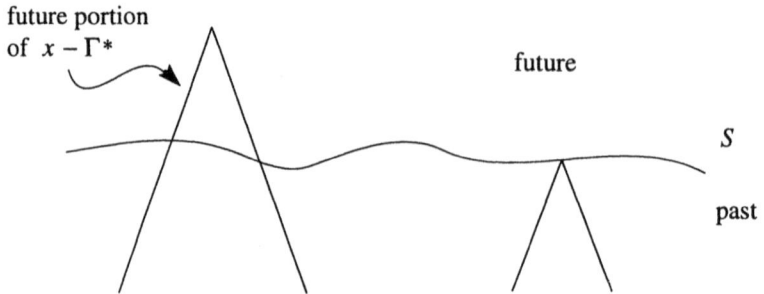

Figure 8.3

only on the values of f in the future portion of $x - \Gamma^*$. More generally, for the full Cauchy problem $Pu = f$, $(\partial/\partial n)^k u \big|_S = g_k$, $k = 0, \ldots, m - 1$, the solution $u(x)$ for x in the future depends only on the values of f in the future portion of $x - \Gamma^*$, and the values of g_k on the portion of S that lies in $x - \Gamma^*$. (A similar statement holds for x in the past, with the past portion of $x + \Gamma^*$ in place of the future portion of $x - \Gamma^*$.) This is the finite speed of propagation property of hyperbolic equations, which was already discussed for the wave equation in section 5.3.

We have described completely the definition of hyperbolicity for constant coefficient equations with no lower order terms. The story in general is too complicated to describe here. However, there is special class of hyperbolic operators, called *strictly hyperbolic*, which has the property that we can ignore lower order terms and which can be defined in a naive frozen coefficients fashion. Recall that for a noncharacteristic direction v, we required that the roots of the polynomial $p(\xi + zv)$ be real. For a strictly hyperbolic operator, we require in addition that the roots be distinct, for ξ not a multiple of v. Since we want to deal with the case of variable coefficient operators, we take the top-order symbol $p_m(x, \xi)$ and we say P is *strictly hyperbolic* at x in the direction v if $p_m(x, v) \neq 0$ and the roots of $p_m(x, \xi + zv)$ are real and distinct for all ξ not a multiple of v. We then define a *strictly hyperbolic operator* P to be one for which there is a smooth vector field $v(x)$ such that P is hyperbolic at x in the direction $v(x)$. It is easy to see that the wave operator is strictly hyperbolic because the two roots $-\tau - k|\xi|$ and $-\tau + k|\xi|$ are distinct if $\xi \neq 0$ (this is exactly the condition that (τ, ξ) not be a multiple of $(1, 0)$). For strictly hyperbolic operators we have a theory very much like what was described above, at least locally. Of course the cone Γ may vary from point to point, and the analog of the light cone is a more complicated geometric object. In particular, we can perturb the wave operator by adding terms of order 1 and 0, even with variable coefficients, without essentially altering what was said above. We can even handle a variable speed wave operator $(\partial^2/\partial t^2) - k(x)^2 \Delta_x$ where $k(x)$ is a smooth function, bounded and nonvanishing.

To say that the Cauchy problem is well-posed means, in addition to existence and uniqueness, that the solution depends continuously on the data. This is expressed in terms of a priori estimates involving L^2 Sobolev spaces. A typical estimate (for strictly hyperbolic P) is

$$\|\psi u\|_{L^2_{k+m}} \leq c(\|f\|_{L^2_k} + \|g_0\|_{L^2_{k+m-1/2}} + \|g_1\|_{L^2_{k+m-3/2}} + \cdots + \|g_{m-1}\|_{L^2_{k+1/2}})$$

where $\psi \in \mathcal{D}(\mathbb{R}^n)$ is a localizing factor. Here, of course, the Sobolev norms for u and f refer to the whole space $\mathbb{R}^n$, while the Sobolev norms for g_j refer to the $(n-1)$-dimensional space S (we have only discussed this for the case of flat S). This estimate shows the continuous dependence on data when applied to the difference of two solutions. If u and v have data that are close together in the appropriate Sobolev spaces, then u and v are close in Sobolev norm. Using the Sobolev inequalities, we can get the conclusion that u is close to v in a pointwise sense, at least locally.

If u is in L^2_{k+m}, then Pu will be in L^2_k, since P involves m derivatives, and so the appearance of the L^2_k norm for f on the right side of the a priori estimate is quite natural. However, the Sobolev norms of the boundary terms seem to be off by $-1/2$. In other words, since $g_j = (\partial/\partial n)^j u\big|_S$, we would expect to land in L^2_{m+k-j}, not $L^2_{m+k-j-1/2}$. But we can explain this discrepency because we are restricting to the hypersurface S. There is a general principle that restriction of L^2 Sobolev spaces reduces the number of derivatives by $1/2$ times the codimension (the codimension of a hypersurface being 1).

To illustrate this principle, consider the restriction of $f \in L^2_s(\mathbb{R}^n)$ to the flat hypersurface $x_n = 0$. We can regard this as a function on $\mathbb{R}^{n-1}$, and we want to show that it is in $L^2_{s-1/2}(\mathbb{R}^{n-1})$ if $s > 1/2$. Write $Rf(x_1, \ldots, x_{n-1}) = f(x_1, \ldots, x_{n-1}, 0)$. By the one-dimensional Fourier inversion formula we have $Rf(x_1, \ldots, x_{n-1}) = \frac{1}{2\pi} \int_{-\infty}^{\infty} \int_{-\infty}^{\infty} f(x) e^{ix_n \xi_n} dx_n d\xi_n$ and from this we obtain $Rf\hat{\ }(\xi_1, \ldots, \xi_{n-1}) = \frac{1}{2\pi} \int_{-\infty}^{\infty} \hat{f}(\xi_1, \ldots, \xi_n) d\xi_n$. So the Fourier transform of the restriction Rf is obtained from the Fourier transform of f by integrating out the extra variable.

Now we apply the Cauchy-Schwartz inequality to the integral for $Rf\hat{\ }$, writing $\hat{f}(\xi) = [\hat{f}(\xi)(1 + |\xi|^2)^{s/2}][(1 + |\xi|^2)^{-s/2}]$. We obtain

$$|Rf\,\hat{}\,(\xi_1,\ldots,\xi_{n-1})|^2$$
$$\leq \frac{1}{(2\pi)^2}\left(\int_{-\infty}^{\infty}|\hat{f}(\xi)|^2(1+|\xi|^s\,d\xi_n\right)\cdot\left(\int_{-\infty}^{\infty}(1+|\xi|^2)^{-s}\,d\xi_n\right).$$

Now a direct computation shows

$$\int_{-\infty}^{\infty}(1+|\xi|^2)^{-s}\,d\xi_n = c(1+\xi_1^2+\cdots+\xi_{n-1}^2)^{-s+1/2}$$

if $s > 1/2$ (this follows by the change of variable $\xi_n \to (1+\xi_1^2+\cdots+\xi_{n-1}^2)^{1/2}\xi_n$). Thus we have

$$(1+\xi_1^2+\cdots+\xi_{n-1}^2)^{s-1/2}|Rf\,\hat{}\,(\xi_1,\ldots,\xi_{n-1})|^2 \leq c\int_{-\infty}^{\infty}|\hat{f}(\xi)|^2(1+|\xi|^2)^s\,d\xi_n$$

and integrating with respect to $\xi_1,\ldots,\xi_{n-1}$ we obtain

$$\int_{\mathbb{R}^{n-1}}(1+\xi_1^2+\cdots+\xi_{n-1}^2)^{s-1/2}|Rf\,\hat{}\,(\xi_1,\ldots,\xi_{n-1})|^2\,d\xi_1\ldots d\xi_{n-1}$$
$$\leq c\int_{\mathbb{R}^n}|\hat{f}(\xi)|^2(1+|\xi|^2)^s\,d\xi_n.$$

This says exactly $\|Rf\|_{L^2_{s-1/2}} \leq c\|f\|_{L^2_s}$. Actually, the pointwise restriction may not initially be well defined. To have f continuous, by the Sobolev inequalities, we would have to assume $s > n/2$. With just $s > 1/2$, neither f nor Rf need be continuous. Nevertheless, the a priori estimate $\|Rf\|_{L^2_{s-1/2}} \leq c\|f\|_{L^2_s}$ shows that the restriction operator can be consistently defined.

8.6   The wave front set

If a function fails to be smooth, we can locate the singularity in space, and we can further analyze the direction of the singularity. The key idea for doing this is the wave front set. This is one place where it is better to think about spaces of dimension greater than one right away; in one dimension there are only two directions, and the ideas do not have immediate intuitive appeal.

For example, consider a function on the plane that is identically one on one side of a smooth curve and zero on the other side.

The function is clearly smooth everywhere except along the curve, where it has a jump discontinuity. If one had to choose a direction for the singularity of the function at a point on the curve, it would seem to be the

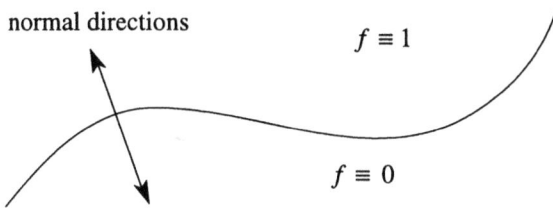

Figure 8.4

normal direction at that point. It would also seem reasonable to say that the function is smooth in the tangential direction. There is less intuitive justification for deciding about all the other directions. We are actually going to give a definition that decides that this function is smooth in all directions except the two normal directions. Here is one explanation for that decision. Choose a direction v, and try to smooth the function out by averaging it in the directions perpendicular to v (in $\mathbb{R}^n$ this will be an $(n-1)$-dimensional space, but for $n = 2$ it is just a line). If you get a smooth function, then the original function must have been smooth in the v direction. Applying this criterion to our example, we see that averaging in any direction other than the tangential direction will smooth out the singularity, and so the function is declared to be smooth in all directions except the normal ones. (Actually, this criterion is a slight oversimplification, because it does not distinguish between v and $-v$.)

Let's consider another example, the function $|x|$. The only singularity is at the origin. Since the function is radial, all directions are equivalent at the origin, so the function cannot be smooth in any direction there.

How do we actually define microlocal smoothness for a function (or distribution)? First we localize in space. We take the function f and multiply by a test function φ supported near the point $\bar{x}$ (we require $\varphi(\bar{x}) \neq 0$ to preserve the behavior of the function f near $\bar{x}$). If f were smooth near $\bar{x}$, then φf would be a C^∞ function of compact support, and so $(\varphi f)\,\hat{}\,(\xi)$ would be rapidly decreasing,

$$|(\varphi f)\,\hat{}\,(\xi)| \leq c_N (1 + |\xi|)^{-N} \text{ for all } N.$$

Then we localize in frequency. There are many different directions to let $\xi \to \infty$. In some directions we may find rapid decrease, in others not. If we

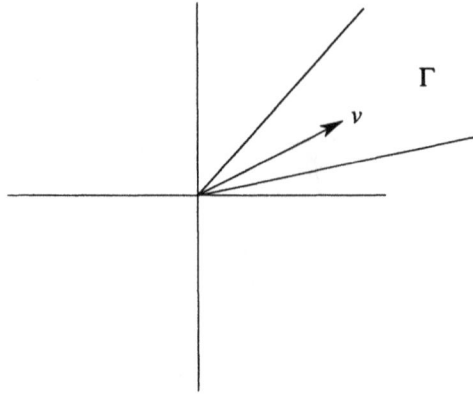

Figure 8.5

find an open cone Γ containing the direction v, such that

$$|(\varphi f)\,\hat{}\,(\xi)| \le c_N (1 + |\xi|)^{-N} \text{ for } \xi \in \Gamma,$$

for all N, then we say f is smooth microlocally at $(\bar{x}, v)$.

More precisely, the order of quantifiers is the following: f is *microlocally smooth at* $(\bar{x}, v)$ *if there exists* $\varphi \in \mathcal{D}$ *with* $\varphi(\bar{x}) \neq 0$ *and an open cone* Γ *containing* v *such that for all* N *there exists* c_N *such that*

$$|(\varphi f)\,\hat{}\,(\xi)| \le c_N (1 + |\xi|)^{-N} \text{ for all } \xi \in \Gamma .$$

The wave front set of f, *denoted* $WF(f)$, *is defined to be the complement of all* $(\bar{x}, v)$ *where* f *is microlocally smooth.*

From the definition it is clear that the points where f is microlocally smooth form an open set, since it contains all (x, ξ) where $\varphi(x) \neq 0$ and $\xi \in \Gamma$. Thus the wave front set is closed. It is also clearly conical in the ξ variable: if $(x, \xi) \in WF(f)$ then also $(x, \lambda\xi) \in WF(f)$ for any $\lambda > 0$. There is one consistency condition that should be checked in the definition. We have to allow φ to be variable, because there might be some singularities of f that are very near $\bar{x}$ that will be cut away by multiplying by φ with very small support, but would not be cut away if the support of φ is too large. However, once we have found a φ that works for showing f is microlocally smooth at $(\bar{x}, v)$, we would like to know that all "smaller" cut-off functions would also work. This is in fact true if we intepret "smaller" to mean $\psi\varphi$

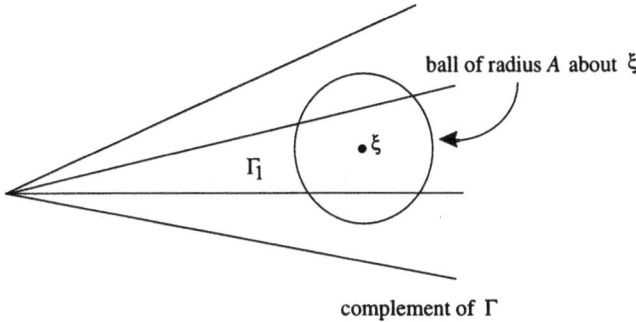

ball of radius A about ξ

Γ_1

$\bullet\xi$

complement of Γ

Figure 8.6

for some other test function ψ. The idea here is to use

$$(\psi\varphi f)\,\hat{}\,(\xi) = \frac{1}{(2\pi)^n}\hat{\psi} * (\varphi f)\,\hat{}\,(\xi).$$

Suppose we know $|(\varphi f)\,\hat{}\,(\xi)| \leq c_N(1+|\xi|)^{-N}$ for all $\xi \in \Gamma$. Then if we take any smaller open cone $\Gamma_1 \subseteq \Gamma$, but still containing v, we will have a similar estimate,

$$|(\psi\varphi f)\,\hat{}\,(\xi)| \leq c'_N(1+|\xi|)^{-N} \text{ for all } \xi \in \Gamma_1.$$

(The constant c'_N in this estimate depends on Γ_1, and blows up if we try to take $\Gamma_1 = \Gamma$.) The "proof by pictures" is to observe that as you go out to infinity in Γ_1, the distance to the complement of Γ also goes to infinity.

In the convolution formula

$$\hat{\psi} * (\varphi f)\,\hat{}\,(\xi) = \int \hat{\psi}(\eta)(\varphi f)\,\hat{}\,(\xi - \eta)\,d\eta$$

the function $\hat{\psi}(\eta)$ is rapidly decreasing in all directions, so it is very close to having compact support. If we knew that $\hat{\psi}(\eta)$ were supported in $|\eta| \leq A$, then for ξ large enough in Γ_1 we would have $\xi - \eta$ in Γ, so we could use the estimate for $(\varphi f)\,\hat{}$ to get

$$|\hat{\psi} * (\varphi f)\,\hat{}\,(\xi)| \leq c\int_{|\eta|\leq A}(1+|\xi - \eta|)^{-N}d\eta$$

and this is dominated by a multiple of $(1+|\xi|)^{-N}$ once $|\xi| \geq 2A$. Since $\hat{\psi}$ does not actually have compact support, the argument is a little more complicated, but this is the essential idea.

Although the definition of the wave front set is complicated, the concept turns out to be very useful. We can already give a typical application. Multiplication of two distributions is not defined in general. We have seen that it can be defined if one or the other factor is a C^∞ function, and it should come as no surprise that it can be defined if the singular supports are disjoint. If T_1 and T_2 are distributions with disjoint singular support, then T_1 is equal to a C^∞ function on the complement of sing supp T_2, so $T_1 \cdot T_2$ is defined there. Similarly $T_2 \cdot T_1$ is defined on the complement of sing supp T_1 because T_2 is equal to a C^∞ function there, and by piecing the two together we get a product defined everywhere. Now I claim that we can even define the product if the singular supports overlap, provided that the wave front sets satisfy a separation condition. The condition we need is the following: we never have both (x, v) in $WF(T_1)$ and $(x, -v)$ in $WF(T_2)$. The occurance of the minus sign will be clear from the proof.

To define $T_1 \cdot T_2$ it suffices to define $\varphi^2 T_1 \cdot T_2$ for test functions φ of sufficiently small support, for then we can piece together $T_1 \cdot T_2 = \sum \varphi_j^2 T_1 \cdot T_2$ from such test functions with $\sum \varphi_j^2 \equiv 1$. Now we will try to define $\varphi^2 T_1 \cdot T_2$ as the inverse Fourier transform of $\frac{1}{(2\pi)^n}(\varphi T_1)\hat{\ } * (\varphi T_2)\hat{\ }$. The problem is that the integral defining the convolution may not make sense. Since φT_1 and φT_2 have compact support, we do know that they satisfy estimates

$$|(\varphi T_1)\hat{\ }(\xi)| \leq c(1 + |\xi|)^{N_1}$$
$$|(\varphi T_2)\hat{\ }(\xi)| \leq c(1 + |\xi|)^{N_2}$$

for some N_1 and N_2. But

$$(\varphi T_1)\hat{\ } * (\varphi T_2)\hat{\ }(\xi) = \int (\varphi T_1)\hat{\ }(\xi - \eta)(\varphi T_2)\hat{\ }(\eta)\, d\eta$$

and these estimates will not make the integral converge. However, for any fixed v, either $(\varphi T_2)\hat{\ }(\eta)$ is rapidly decreasing in an open cone Γ containing v, or $(\varphi T_1)\hat{\ }(-\eta)$ is rapidly decreasing in Γ, if φ is taken with small enough support, by the separation condition on the wave front sets. This is good enough to make the portion of the integral over Γ converge absolutely (for every ξ). In the first alternative

$$\int_\Gamma |(\varphi T_1)\hat{\ }(\xi - \eta)(\varphi T_2)\hat{\ }(\eta)|\, d\eta \leq c_N \int_\Gamma (1 + |\xi - \eta|)^{N_1}(1 + |\eta|)^{-N} d\eta$$

for any N, which will converge if $N_1 - N < -n$; in the second alternative we need the observation that for fixed ξ, $\xi - \eta$ will belong to $-\Gamma$ if η is in a slightly smaller cone Γ_1, for η large enough, so

$$\int_{\Gamma_1} |(\varphi T_1)\hat{\ }(\xi - \eta)(\varphi T_2)\hat{\ }(\eta)|\, d\eta \leq c_N \int_{\Gamma_1} (1 + |\xi - \eta|)^{-N}(1 + |\eta|)^{N_2}\, d\eta,$$

and this converges if $N_2 - N < -n$. By piecing together these estimates, using a compactness argument to show that we can cover $\mathbb{R}^n$ by a finite collection of such cones, we see that the convolution is well defined. These estimates also show that $(\varphi T_1)\,\hat{}\; * (\varphi T_2)\,\hat{}\;$ is of polynomial growth in ξ, so the inverse Fourier transform is also defined.

A more careful examination of the argument shows that the wave front set of the product $T_1 \cdot T_2$ must lie in the union of three sets: the wave front sets of T_1 and T_2, and the set of $(x, \xi_1 + \xi_2)$ where $(x, \xi_1) \in WF(T_1)$ and $(x, \xi_2) \in WF(T_2)$.

We will have more applications of the wave front set shortly, but now let's go through the computation of the wave front set for the examples we discussed from an intuitive point of view before. To be specific, let $f(x, y) = \chi(y > 0)$ be the characteristic function of the upper half-plane. For any point not on the x-axis, we can choose φ equal to one at the point and supported in the upper or lower half-plane, so φf is C^∞. For a point $(x_0, 0)$ on the x-axis, choose $\varphi(x, y)$ of the form $\varphi_1(x)\varphi_2(y)$ where $\varphi_1(x_0) = 1$ and $\varphi_2(0) = 1$. Then $f(x, y)\varphi(x, y) = \varphi_1(x)h(y)$ where

$$h(y) = \begin{cases} \varphi_2(y) & \text{if } y > 0 \\ 0 & \text{if } y \leq 0. \end{cases}$$

(Since we are free to choose φ rather arbitrarily, we make a choice that will simplify the computations.) The Fourier transform of $\varphi \cdot f$ is then $\hat{\varphi}_1(\xi)\hat{h}(\eta)$. Now $\hat{\varphi}_1(\xi)$ is rapidly decreasing, but $\hat{h}(\eta)$ is not, because h has a jump discontinuity. (In fact, $\hat{h}(\eta)$ decays exactly at the rate $O(|\eta|^{-1})$, but this is not significant.) Now suppose we choose a direction $v = (v_1, v_2)$ and let (ξ, η) go to infinity in this direction; say $(\xi, \eta) = (tv_1, tv_2)$ with $t \to +\infty$. Then the Fourier transform restricted to this ray will be $\hat{\varphi}_1(tv_1)\hat{h}(tv_2)$. As long as $v_1 \neq 0$, the rapid decay of $\hat{\varphi}_1$ will make this product decay rapidly (for this it is enough to have $|\hat{h}(\eta)| \leq c(1 + |\eta|)^N$ for some N, which holds automatically since h has compact support). But if $v_1 = 0$ then $\hat{\varphi}_1$ does not contribute any decay and $\hat{h}(tv_2)$ does not decay rapidly. This shows that f is microlocally smooth in all directions except the y-axis, so $WF(f)$ consists of the pairs $((x, 0), (0, v_2))$, the boundary curve, and its normal directions. This agrees with our intuitive guess.

Of course we took a very simple special case, where the boundary curve was a straight line, and one of the axes in addition. For a more general curve the computations are more difficult, and we prefer a roundabout approach. Any smooth curve can be straightened out by a change of coordinates, so we would like to know what happens to the wave front set under such a mapping. This is an important question in its own right, because the answer will help explain what kind of space the wave front set lies in.

Suppose we apply a linear change of variable $y = Ax$ where A is an invertible $n \times n$ matrix. Recall that if T is a distribution we defined $T \circ A$ as a distribution by

$$\langle T \circ A, \varphi \rangle = |\det A|^{-1} \langle T, \varphi \circ A^{-1} \rangle$$

and if T is a distribution of compact support then

$$
\begin{aligned}
(T \circ A)\,\hat{}\,(\xi) &= |\det A|^{-1} \langle T, e^{i(A^{-1}x)\cdot\xi} \rangle \\
&= |\det A|^{-1} \hat{T}((A^{-1})^{tr}\xi)
\end{aligned}
$$

because $A^{-1}x \cdot \xi = x \cdot (A^{-1})^{tr}\xi$. Now fix a point $\bar{x}$ and a direction ξ. If we multiply $T \circ A$ by φ where $\varphi(\bar{x}) \neq 0$, this is the same as $(\psi T) \circ A$ where $\psi = \varphi \circ A^{-1}$ and $\psi(\bar{y}) = \varphi(A^{-1}\bar{y}) = \varphi(\bar{x}) \neq 0$ where $\bar{y} = A\bar{x}$. Then

$$(\varphi(T \circ A))\,\hat{}\,(t\xi) = ((\psi T) \circ A)\,\hat{}\,(t\xi) = |\det A|^{-1}(\psi T)\,\hat{}\,(t(A^{-1})^{tr}\xi)$$

so $(\bar{x}, \xi) \in WF(T \circ A)$ if and only if $(A\bar{x}, (A^{-1})^{tr}\xi) \in WF(T)$. Note that if A is a rotation, then $(A^{-1})^{tr} = A$ and so the same rotation acts on both variables. Also a rotation preserves perpendicularity, so our conclusion that the wave front set of the characteristic function of the upper half-plane points perpendicular to the boundary carries over to any half-plane. But to understand regions bounded by curves we need to consider nonlinear coordinate changes.

Let $g : \mathbb{R}^n \to \mathbb{R}^n$ be a smooth mapping with smooth inverse g^{-1}. Then for any distribution T we may define $T \circ g$ by

$$\langle T \circ g, \varphi \rangle = \langle T, J(g)^{-1}\varphi \circ g^{-1} \rangle$$

where $J(g) = |\det g'|$ is the Jacobian. The analogous statement for the wave front set is the following: $(x, \xi) \in WF(T \circ g)$ if and only if $(g(x), (g'(x)^{-1})^{tr}\xi) \in WF(T)$. Note that this is consistent with the linear case because a linear map is equal to its derivative.

This transformation law for wave front sets preserves perpendicularity to a curve. Suppose $x(t)$ is a curve and $(x(t_0), \xi)$ is a point in $WF(f \circ g)$ with ξ perpendicular to the curve, $x'(t_0)\cdot\xi = 0$. Then $(g(x(t_0)), (g'(x(t_0))^{-1})^{tr}\xi)$ is a point in $WF(f)$ with $(g'(x(t_0))^{-1})^{tr}\xi$ perpendicular to the curve $g(x(t))$. This follows because $g'(x(t))x'(t)$ is the tangent to the curve at $g(x(t_0))$, and

$$
\begin{aligned}
g'(x(t_0))x'(t_0) \cdot (g'(x(t_0))^{-1})^{tr}\xi \\
= (g'(x(t_0))^{-1}g'(x(t_0))x'(t_0)) \cdot \xi \\
= x'(t_0) \cdot \xi = 0.
\end{aligned}
$$

So if f is the characteristic function of a region bounded by a curve $\gamma(t)$, we choose g to map the upper half-plane to this region, and the x-axis to the curve $\gamma(t)$. Since we know that $WF(f \circ g)$ consists of pairs (x, ξ) where x lies on the x-axis and ξ is perpendicular to it, it follows that $WF(f)$ consists of pairs (x, ξ) where x lies on the curve $\gamma(t)$ and ξ is perpendicular to it.

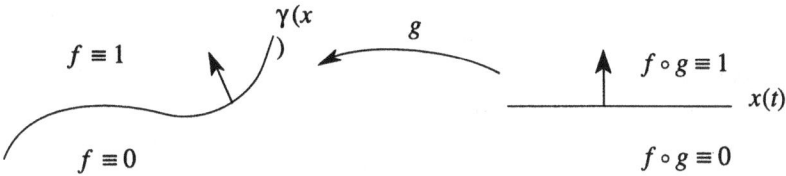

Figure 8.7

The transformation property of the wave front set indicates that the second variable ξ should be thought of as a *cotangent* vector, not a tangent vector. In the more general context of manifolds, the wave front set is a subset of the cotangent bundle. The distinction between tangents and cotangents is subtle, but worth understanding. Tangents may be thought of as directions, or arrows. The tangents at a fixed point (the base of the arrows) form a vector space. (Of course if we are working in Euclidean space then the space is already a vector space, so it is better to think about a surface, say the unit sphere in $\mathbb{R}^3$. At every point on the sphere, the tangent space is a two-dimensional vector space of arrows tangent to the sphere at that point). The cotangent space is the dual vector space. Although a finite-dimensional vector space may be identified with its dual, this identification is not canonical, but requires the choice of an inner product (in the manifold context this is a Riemannian metric). In our case, the ξ gives rise to a linear function on tangent directions $v \to v \cdot \zeta$, but is not itself a tangent direction. This explains why the intuitive motivation for the definition of wave front set in terms of singularities in "directions" is so complicated. The "directions" are not really tangent directions after all. (The same is true of characteristic "directions" for differential operators, by the way.) However, it does make sense to talk about the subspace of directions perpendicular to ξ (in $\mathbb{R}^2$ this is a one-dimensional space), since this is defined by $v \cdot \xi = 0$.

Now we consider another application of the wave front set, this time to the question of restrictions of distributions to lower dimensional surfaces.

This is even a problem for functions. In the examples we have been considering of a function with a jump discontinuity along a curve, it does not make sense to restrict the function to the curve, since its value is in some sense undefined along the whole curve. However, it does make sense to restrict it to a different curve that cuts across it at an angle, for then there is only one ambiguous point along the curve. *In fact, it turns out that restrictions can be defined, even for distributions, as long as the normal directions to the surface are not in the wave front set of the distribution.*

We illustrate this principle in the special case of restrictions from the plane to the x-axis. Suppose T is a distribution on $\mathbb{R}^2$ whose wave front set does not contain any pair $((x,0),(0,t))$ consisting of a point on the x-axis and a cotangent perpendicular to it. Then we claim there is a natural way to define the restriction $T(x,0)$ as a distribution on the line. It suffices to define the restriction of φT for φ a test function of small support, for then we can piece together the restrictions of $\varphi_j T$ where $\Sigma \varphi_j \equiv 1$ and get the restriction of T. Now for any point on the x-axis, if we take the support of φ in a small enough neighborhood of the point, then $(\varphi T)\hat{\ }$ will be rapidly decreasing in an open cone Γ containing the vector $(0,1)$, and similarly for $-\Gamma$ containing $(0,-1)$, since by assumption T is microlocally smooth in these cotangent directions. To define the restriction of φT we will define its Fourier transform. But we have seen that the Fourier transform of the restriction should be given by integrating the Fourier transform of φT. In other words, if R denotes the restriction, we should have

$$R(\varphi T)\hat{\ }(\xi) = \frac{1}{2\pi} \int_{-\infty}^{\infty} (\varphi T)\hat{\ }(\xi, \eta)\, d\eta.$$

What we claim is that this integral converges absolutely. This will enable us to define $R(\varphi T)\hat{\ }$ by this formula, and that is the definition of $R(\varphi T)$ as a distribution on $\mathbb{R}^1$. The proof of the convergence of the integal follows from the picture that shows that for any fixed ξ, the vertical line (ξ, η) eventually lies in Γ or $-\Gamma$ as $\eta \to \pm\infty$.

Once inside Γ or $-\Gamma$, the rapid decay guarantees the convergence, and the finite part of the line not in these two cones just produces a finite contribution to the integral, since $(\varphi T)\hat{\ }$ is continuous. A more careful estimate shows that $R(\varphi T)\hat{\ }(\xi)$ has polynomial growth, so the inverse Fourier transform in $\mathcal{S}'$ is well defined.

To prove the restriction property for a general curve in the plane, we simply apply a change of coordinates to straighten out the curve. The hypothesis that the normal cotangents not be in the wave front set is preserved by the transformation property of wave front sets, as we have already observed. A similar argument works for the restriction from $\mathbb{R}^n$ to any smooth k-dimensional surface. Note that in this case the normal direc-

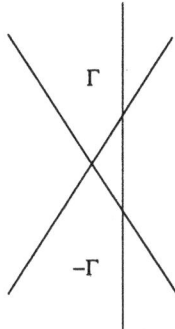

Figure 8.8

tions to a point on the surface form a subspace of dimension $n - k$ of the cotangent space.

A more refined version of this argument allows us to locate the wave front set of the restriction. Suppose $\bar{x}$ is a point on the surface S to which we restrict T. If we take the cotangents ξ at $\bar{x}$ for which $(\bar{x}, \xi) \in WF(T)$ and project them orthogonally onto the cotangent directions to S, then we get the cotangents that may lie in the wave front set of the restriction of T to S at $\bar{x}$. In other words, if we have a cotangent η to S at $\bar{x}$ that is not the projection of any cotangent in the wave front set of T of $\bar{x}$, then RT is microlocally smooth at $(\bar{x}, \eta)$.

8.7 Microlocal analysis of singularities

Where do the singularities go when you solve a differential equation? We can now begin to answer this question on a microlocal level. The first observation is that the wave front set is not increased by applying a differential operator,

$$WF(Pu) \subseteq WF(u).$$

This is analogous to the fact that P is a local operator (the singular support is not increased), and so we say that any operator with this property is a *microlocal operator*. In fact it is true that pseudodifferential operators are also microlocal.

We illustrate this principle by considering the case of a constant coefficient differential operator. The microlocal property is equivalent to the

statement that if u is microlocally smooth at (x, ξ) then so is Pu. Now
the idea is that if $(\varphi u)\hat{\ }$ is rapidly decreasing in an open cone Γ, then so
is $P(\varphi u)\hat{\ }$ since this is just the product of $(\varphi u)\hat{\ }$ with the polynomial p.
This is almost the entire story, except that what we have to show is that
$(\varphi Pu)\hat{\ }$ is rapidly decreasing in Γ, and $\varphi Pu \neq P(\varphi u)$. In this case this is
just a minor inconvenience, since $\varphi Pu = P(\varphi u) + \Sigma \varphi_j Q_j u$ where Q_j are
lower order constant coefficient differential operators and φ_j are test functions with the same support as φ. Thus we prove the assertion by induction
on the order of the differential operator, and the induction hypothesis takes
care of the $\varphi_j Q_j u$ terms.

Since applying a differential operator does not increase the wave front
set, the next natural question to ask is, can it decrease it? For elliptic
operators, the answer is no. This is the microlocal analog of hypoellipticity
and so is called *microlocal hypoellipticity*. It is easy to understand this fact
in terms of the parametrix. If P is an elliptic operator and Q is the ψDO
parametrix, so $QPu = u + Ru$ with Ru a C^∞ function, then the microlocal
nature of Q implies

$$WF(u) = WF(QPu) \subseteq WF(Pu).$$

So for elliptic operators $WF(Pu) = WF(u)$. For nonelliptic operators
we can say where $WF(u)$ may be located, in terms of the characteristics
of P. Recall that we said the cotangent ξ is characteristic for P at x if
$p_m(x, \xi) = 0$, where p_m is the top-order symbol of P. Let's call char(P)
the set of all pairs (x, ξ) where this happens.

Theorem 8.7.1 $WF(u) \subseteq WF(Pu) \cup \text{char}(P)$. *Another way of stating
this theorem is that if Pu is microlocally smooth at (x, ξ), and ξ is noncharacteristic at x, then u is microlocally smooth at (x, ξ).*

The proof of this fact is just a variation on the construction of a para -
metrix for an elliptic operator. In this case we construct a ψDO of order $-m$
which serves as a microlocal parametrix, so QP is not equal to the identity
plus R, but rather $QP = L + R$ where L is a microlocalizer, restricting to
a neighborhood of the point x and an open cone Γ containing ξ. By its
construction, Lu and u will have the same microlocal behavior at (x, ξ), Ru
is always smooth, and QPu is microlocally smooth at (x, ξ) because of the
microlocal property of Q. In constructing Q we simply use the nonvanishing
and the homogeneity of p_m in a microlocal neighborhood of (x, ξ).

We can use this theorem to make sense of the boundary values in the
Cauchy problem. If S is a noncharacteristic surface for P, then none of the
normal directions to S are contained in char(P). If $Pu = 0$ (or $Pu = f$ for
smooth f) then $WF(u) \subseteq \text{char}(P)$ by the theorem. Thus $WF(u)$ contains

none of the normals to S, so the restriction to S is well defined. Since the wave front set of any derivative of u is contained in $WF(u)$, the same argument shows that boundary values of derivatives of any order are also well defined.

We are now going to consider a more refined statement about the singularities of solutions of $Pu = 0$ (or $Pu = f$ with smooth f). We know they lie in $\mathrm{char}(P)$, but we would like to be able to say more, since $\mathrm{char}(P)$ may be a very large set. We will break $\mathrm{char}(P)$ up into a disjoint union of curves called *bicharacteristics*. Then we will have an all-or-nothing dichotomy: either $WF(u)$ contains the whole bicharacteristic or no points of it. This is paraphrased by saying *singularities propagate along bicharacteristics*. To do this we have restrict attention to a class of operators called operators of *principal type*. This is a rather broad class of operators that includes both elliptic operators and strictly hyperbolic operators. The definition of principal type is that the characteristics be simple zeroes of the top-order symbol: if $p_m(x, \xi) = 0$ for $\xi \neq 0$ then $\nabla_\xi p_m(x, \xi) \neq 0$. Note that this definition depends only on the highest order terms, and it has a "frozen coefficient" nature in the dependence on x. The intuition is that operators of principal type have the property that the highest order terms "dominate" the lower order terms. We will also assume that the operator has real valued coefficients, so $i^m p_m(x, \xi)$ is a real valued function.

The definition of *bicharacteristic* in general (for operators with real valued coefficients) is a curve $(x(t), \xi(t))$ which satisfies the Hamilton equations

$$\frac{dx_j(t)}{dt} = i^m \frac{\partial p_m}{\partial \xi_j}(x(t), \xi(t)) \qquad j = 1, \ldots, n$$

$$\frac{dx_j(t)}{dt} = -i^m \frac{\partial p_m}{\partial x_j}(x(t), \xi(t)) \qquad j = 1, \ldots, n.$$

It is important to realize that this is a system of first-order ordinary differential equations in the $2n$ unknown functions $x_j(t)$ and $\xi_j(t)$ (the partial derivatives on the right are applied to the known function $p_m(x, \xi)$). Therefore, the existence and uniqueness theorem for ordinary differential equations implies that there is a unique solution for every initial value of x and ξ. Now a simple computation using the chain rule shows that p_m is constant along a bicharacteristic curve, because

$$\frac{d}{dt} p_m(x, (t), \xi(t)) = \sum_{j=1}^{n} \left(\frac{\partial p_m}{\partial x_j}(x(t), \xi(t)) \frac{dx_j(t)}{dt} \right.$$

$$\left. + \frac{\partial p_m}{\partial \xi_j}(x(t), \xi(t)) \frac{d\xi_j(t)}{dt} \right)$$

$$= (-i)^m \sum_{j=1}^{n} \left(-\frac{d\xi_j(t)}{dt}\frac{dx_j(t)}{dt} + \frac{dx_j(t)}{dt}\frac{d\xi_j(t)}{dt} \right) = 0.$$

We will only be interested in the *null bicharacteristics*, which are those for which p_m is zero. Thus the set char(P) splits into a union of null bicharacteristic curves. Note that the assumption that P is of principal type means that the x-velocity is nonzero along null bicharacteristic curves. This guarantees that they really are curves (they do not degenerate to a point), and the projection onto the x coordinates is also a curve. We will follow the standard convention in deleting the adjective "null"; all our bicharacteristic curves will be null bicharacteristic curves.

If P is a constant coefficient operator, then $\partial p_m/\partial x_j \equiv 0$ and so the bicharacteristic curves have a constant ξ value. Also $\partial p_m/\partial \xi_j$ is independent of x, and so $dx_j(t)/dt = (\partial p_m/\partial \xi_j)(\xi)$ is a constant; hence $x(t)$ is a straight line in the direction $\nabla p_m(\xi)$. We call these projections of bicharacteristic curves *light rays*.

For example, suppose P is the wave operator $(\partial^2/\partial t^2) - k^2\Delta x$. Then $p_m(\tau, \xi) = -\tau^2 + k^2|\xi|^2$, and the characteristics are $\tau = \pm k|\xi|$, or in other words the cotangents $(\pm k|\xi|, \xi)$ for $\xi \neq 0$. Starting at a point $(\bar{t}, \bar{x})$, and choosing a characteristic cotangent, the corresponding light ray (with parameter s) is given by $t = \bar{t} \pm 2k|\xi|s$ and $x = \bar{x} + 2k^2 s\xi$. If we use the first equation to eliminate the parameter s and reparametrize the light ray in terms of t, the result is

$$x = \bar{x} \pm \frac{k\xi}{|\xi|}(t - \bar{t})$$

which is in fact a light ray traveling at speed k in the direction $\pm\xi/|\xi|$.

Now we can state the propagation of singularities principle.

Theorem 8.7.2 *Let P be an operator of principal type, with real coefficients, and suppose $Pu = f$ is C^∞. For each null bicharacteristic curve, either all points are in the wave front set of u or none are.*

Let's apply this theorem to the wave equation. Suppose we know the solution in a neighborhood of $(\bar{t}, \bar{x})$ and we can compute which cotangents $(\pm k|\xi|, \xi)$ are in the wave front set at that point. Then we know the singularity will propagate along the light rays in the $\pm\xi/|\xi|$ direction passing through that point. If $(\pm k|\xi|, \xi)$ is a direction of microlocal smoothness, then it will remain so along the entire light ray. Thus the "direction" of the singularity dictates the direction it will travel.

In dealing with the wave equation, it is usually more convenient to describe the singularities in terms of Cauchy data. Suppose we are given

$u(x, 0) = f(x)$ and $u_t(x, 0) = g(x)$ for u a solution of

$$\left(\frac{\partial^2}{\partial t^2} - k^2 \Delta_x \right) u = 0.$$

What can we say about the singularities of u at a later time in terms of the singularities of f and g. Fix a point $\bar{x}$ in space $\mathbb{R}^n$, and consider the wave front set of u at the point $(0, \bar{x})$ in spacetime $\mathbb{R}^{n+1}$. We know that the cotangents (τ, ξ) must be characteristics of the equation, so must be of the form $(\pm k|\xi|, \xi)$. Note that exactly two of these characteristic spacetime cotangents project onto each space cotangent ξ. So if $(\bar{x}, \xi)$ belongs to $WF(f)$ or $WF(g)$, then either $((0, \bar{x}), (k|\xi|, \xi))$ or $((0, \bar{x}), (-k|\xi|, \xi))$ belongs to $WF(u)$ (usually both, although it is possible to construct examples where only one does). Conversely, if both f and g are microlocally smooth at $(\bar{x}, \xi)$, then it can be shown that u is microlocally smooth at both points $((0, \bar{x}), (\pm k|\xi|, \xi))$.

Now fix a point in spacetime $(\bar{t}, \bar{x})$. Given a characteristic cotangent $(\pm k|\xi|, \xi)$, we want to know if u is microlocally smooth there. The light ray passing through $(\bar{t}, \bar{x})$ corresponding to this cotangent intersects the $t = 0$ subspace at the point $\bar{x} \mp k\bar{t}(\xi/|\xi|)$. So if f and g are mircolocally smooth in the direction ξ at this point, we know that u is microlocally smooth along the whole bicharacteristic curve, hence at $((\bar{t}, \bar{x}), (\pm k|\xi|, \xi))$. In particular, if we want to know that u is smooth in a neighborhood of $(\bar{t}, \bar{x})$, we have to show that f and g are microlocally smooth at all points $(\bar{x} \mp k\bar{t}(\xi/|\xi|), \xi)$. This requires that we examine the sphere of radius $k\bar{t}$ about $\bar{x}$, but it does *not* require that f and g have no singularities on this sphere; it just requires that the singularities not lie in certain directions.

To illustrate this phenomenon further, let's look at a specific choice of Cauchy data. Let γ be a smooth closed curve in the plane, and let f be the characteristic function of the interior of γ. Also take $g = 0$. So the singular support of f is γ, and the wave front set consists of $(\bar{x}, \xi)$ where $\bar{x}$ is in γ at ξ is normal to γ. This leads to singularities at a later time $\bar{t}$ at the points $\bar{x} \pm k\bar{t}(\xi/|\xi|)$.

Clearly it suffices to consider ξ of unit length, and in view of the choice of sign we may restrict ξ to be the unit normal pointing outward. Thus the singularities at the time $\bar{t}$ lie on the two curves $\bar{x} \pm k\bar{t}n$ where n denotes this unit normal vector.

In other words, starting from the curve γ, we travel in the perpendicular direction a distance $k\bar{t}$ in both directions to get the two new curves. Furthermore, at each point on the two new curves, the wave front set of $u(\bar{t}, x)$ as a function of x is exactly in the n direction.

Incidentally, we can verify in this situation that the direction n is still perpendicular to the new curves. If $x(s)$ traces out γ with parameter s,

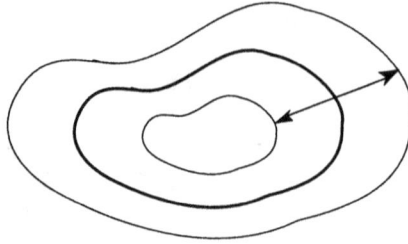

Figure 8.9

then $x'(s)$ is the tangent direction so $n(s)$ satisfies $x'(s) \cdot n(s) = 0$. Also we have $n(s) \cdot n(s) = 1$ because it is of unit length. Differentiating this identity we obtain $n'(s) \cdot n(s) = 0$. But now the new curves are parametrized by s as $x(s) \pm k\bar{t}n(s)$ and so their tangent directions are given by $x'(s) \pm k\bar{t}n'(s)$. Since $(x'(s) \pm k\bar{t}n'(s)) \cdot n(s) = x'(s) \cdot n(s) \pm k\bar{t}n'(s) \cdot n(s) = 0$ we see that $n(s)$ is still normal to the new curves.

8.8 Problems

1. Show that if $f \in L^{p_1}$ and $f \in L^{p_2}$ with $p_1 < p_2$, then $f \in L^p$ for $p_1 \leq p \leq p_2$. (*Hint*: Split f in two pieces according to whether or not $|f| \leq 1$.)

2. Let
$$f(x) = \begin{cases} |\log|x||^\alpha & |x| \leq 1 \\ 0 & |x| \geq 1 \end{cases}$$
   for $0 < \alpha < 1/2$. Show that $f \in L_1^2(\mathbb{R}^2)$ but f is unbounded.

3. If $f \in L_2^2(\mathbb{R}^2)$, show that f satisfies the Zygmund class estimate
$$|f(x+2y) - 2f(x+y) + f(x)| \leq c|y|.$$

4. If $f_1 \in L_k^2(\mathbb{R}^1)$ and $f_2 \in L_k^2(\mathbb{R}^1)$, show that $f_1(x_1)f_2(x_2) \in L_k^2(\mathbb{R}^2)$.

5. If $f \in L_1^2(\mathbb{R}^2)$ show that $(\partial/\partial r)f \in L^2(\mathbb{R}^2)$. Give an example to show $(\partial/\partial\theta)f$ is not necessarily in $L^2(\mathbb{R}^2)$.

6. For which negative values of s does $\delta \in L_s^2(\mathbb{R}^n)$?

7. Show that $f \in L_2^2(\mathbb{R}^n)$ if and only if $f \in L^2(\mathbb{R}^n)$ and $\Delta f \in L^2(\mathbb{R}^n)$. If $f \in L^2(\mathbb{R}^2)$ and $\partial f/\partial x_1 \partial x_2 \in L^2(\mathbb{R}^2)$, does it necessarily follow that $f \in L_2^2(\mathbb{R}^2)$? (*Hint*: Work on the Fourier transform side.)

8. If $0 < s < 2$, show that $f \in L^2_s(\mathbb{R}^n)$ if and only if $f \in L^2(\mathbb{R}^n)$ and $\int_{\mathbb{R}^n} \int_{\mathbb{R}^n} |f(x+2y) - 2f(x+y) + f(x)|^2 |y|^{-n-2s} dy\, dx < \infty$.

9. For which values of p and k does $e^{-|x|}$ belong to $L^p_k(\mathbb{R}^n)$?

10. If $f \in L^p_k(\mathbb{R}^n)$ and $\varphi \in \mathcal{D}(\mathbb{R}^n)$, show that $\varphi \cdot f \in L^p_k(\mathbb{R}^n)$.

11. Show that the heat operator $(\partial/\partial t) - k\Delta x$ is not elliptic, but its full symbol has no real zeroes except the origin. Show that it satisfies the a priori estimate

$$\|u\|_{L^2_{k+1}} \leq c(\|u\|_{L^2} + \|Pu\|_{L^2_k}).$$

12. Show that the heat operator is hypoelliptic by showing

$$\left| \left(\frac{\partial^2}{\partial \tau^2} + \Delta_\xi \right)^N p(\tau, \xi)^{-1} \right| \leq c(\tau^2 + |\xi|^2)^{-\left(\frac{N+1}{2}\right)}$$

for $\tau^2 + |\xi|^2 \geq 1$.

13. Show that if P is any linear differential operator for order m, then

$$P(uv) = \sum_{|\alpha| \leq m} \frac{1}{\alpha!} \left(\left(\frac{\partial}{\partial x} \right)^\alpha u \right) P^{(\alpha)} v$$

where $P^{(\alpha)}$ denotes the differential operator whose symbol is

$$\left(-i \frac{\partial}{\partial \xi} \right)^\alpha p(x, \xi).$$

14. Show that

$$\sum_{j=1}^n \left(\frac{\partial}{\partial x_j} \right)^m$$

is an elliptic operator on $\mathbb{R}^n$ if m is even, but not if m is odd.

15. Show that $\Delta u = \lambda u$ has no solutions of polynomial growth if $\lambda > 0$, but does have such solutions if $\lambda < 0$.

16. Let $P_1, \ldots, P_k$ be first-order differential operators on $\mathbb{R}^n$. Show that $P_1^2 + \cdots + P_k^2$ is never elliptic if $k < n$.

17. Show that a first-order differential operator on $\mathbb{R}^n$ cannot be elliptic if either $n \geq 3$ or the coefficients are real-valued.

18. Show that a C^∞ function that is homogeneous of any degree must be a polynomial. (*Hint*: A high enough derivative would be homogeneous of negative degree, hence not continuous at the origin.)

19. Show that the Hilbert transform $Hf = \mathcal{F}^{-1}(\operatorname{sgn}\xi\hat{f}(\xi))$ is a ψDO of order zero in $\mathbb{R}^1$. Show the same for the Riesz transforms

$$R_j f = \mathcal{F}^{-1}\left(\frac{\xi_j}{|\xi|}\hat{f}(\xi)\right)$$

in $\mathbb{R}^n$.

20. Verify that the asymptotic formula for the symbol of a composition of ψDO's gives the exact symbol of the composition of two partial differential operators.

21. Verify that the commutator of two partial differential operators of orders m_1 and m_2 is of order $m_1 + m_2 - 1$.

22. If P is a ψDO of order $r < -n$ with symbol $p(x,\xi)$, show that $k(x,y) = \frac{1}{(2\pi)^n}\int p(x,\xi)e^{-i(x-y)\cdot\xi}d\xi$ is a convergent integral and $Pu(x) = \int k(x,y)u(y)\,dy$ if u is a test function.

23. If $\sigma(x,\xi)$ is a classical symbol of order r and $h : \mathbb{R}^n \to \mathbb{R}^n$ is a C^∞ mapping with C^∞ inverse, show that $\sigma(h(x), (h'(x)^{tr})^{-1}\xi)$ is also a classical symbol of order r.

24. Show that $\sigma(x,\xi) = (1 + |\xi|^2)^{\alpha/2}$ is a classical symbol of order α. (*Hint*: Write $(1 + |\xi|^2)^{\alpha/2} = |\xi|^\alpha(1 + |\xi|^{-2})^{\alpha/2}$ and use the binomial expansion for $|\xi| > 1$.)

25. If P is a ψDO of order zero with symbol $p(\xi)$ that is independent of x, show that $||Pu||_{L^2} \leq ||u||_{L^2}$. (*Hint*: Use the Plancherel formula.)

26. If h is a C^∞ function, show that $h\Delta$ is elliptic if and only if h is never zero.

27. Find a fundamental solution for the Hermite operator $-(d^2/dx^2) + x^2$ in $\mathbb{R}^1$ using Hermite expansions. Do the same for $-\Delta + |x|^2$ in $\mathbb{R}^n$.

28. Let P be any second-order elliptic operator with real coefficients on $\mathbb{R}^n$. Show that $(\partial^2/\partial t^2) - P$ is strictly hyperbolic in the t direction on $\mathbb{R}^{n+1}$.

29. In which directions is the operator $\partial^2/\partial x\partial y$ hyperbolic in $\mathbb{R}^2$?

30. Show that the surface $y = 0$ in $\mathbb{R}^3$ is noncharacteristic for the wave operator

$$\frac{\partial^2}{\partial t^2} - \frac{\partial^2}{\partial x^2} - \frac{\partial^2}{\partial y^2}$$

but is not spacelike.

31. For a smooth surface of the form $t = h(x)$, what are the conditions on h in order that the surface be spacelike for the variable speed wave operator $(\partial^2/\partial t^2) - k(x)^2 \Delta_x$?

32. Show that if $f \in L_s^2(\mathbb{R}^n)$ with $s > 1$ then $Rf(x_1, \ldots, x_{n-2}) = f(x_1, \ldots, x_{n-2}, 0, 0)$ is well defined in $L_{s-1}^p(\mathbb{R}^{n-2})$. (*Hint*: Interate the codimension-one restrictions.)

33. Show that if $n \geq 2$, no operator can be both elliptic and hyperbolic.

34. Show that the equation $Pu = 0$ can never have a solution of compact support for P a constant coefficient differential operator. (*Hint*: What would that say about $\hat{u}$?)

35. If u is a distribution solution of $\partial^2 u/\partial x \partial y = 0$ and v is a distribution solution of

$$\frac{\partial^2 v}{\partial x^2} - \frac{\partial^2 v}{\partial y^2} = 0$$

in $\mathbb{R}^2$, show that the product is uv is well defined.

36. For the anisotripic wave operator

$$\frac{\partial^2}{\partial t^2} - k_1^2 \frac{\partial^2}{\partial x_1^2} - k_2^2 \frac{\partial^2}{\partial x_2^2}$$

in $\mathbb{R}^2$, compute the characteristics, the light cone, the bicharacteristic curves and the light rays.

37. Show that the operator

$$\left(\frac{\partial^2}{\partial x^2} - a^2 \frac{\partial^2}{\partial y^2} \right) \left(\frac{\partial^2}{\partial x^2} - b^2 \frac{\partial^2}{\partial y^2} \right)$$

in $\mathbb{R}^2$ is of principal type if and only if $a^2 \neq b^2$.

38. Show that the operator

$$\sum_{j=1}^{n} a_j(x) \frac{\partial^2}{\partial r_j^2}$$

in $\mathbb{R}^2$ is of principal type if the real functions $a_j(x)$ are never zero. When is it elliptic?

39. Let $P = a(x,y)\frac{\partial^2}{\partial x} + b(x,y)\frac{\partial^2}{\partial y}$ be a first-order operator on $\mathbb{R}^2$ with real coefficients, and assume that the coefficients $a(x,y)$ and $b(x,y)$ do not both vanish at any point. Compute the characteristic and the bicharacteristic curves, and show that the light rays are solutions of the system $x'(t) = a(x(t), y(t)), y'(t) = b(x(t), y(t))$.

Suggestions for Further Reading

For a more thorough and rigorous treatment of distribution theory and Fourier transforms:

"Generalized functions, vol. I" by I.M. Gelfand and G.E. Shilov, Academic Press, 1964.

For a deeper discussion of Fourier analysis:

"An introduction to Fourier series and integrals" by R.T. Seeley, Benjamin, 1966.
"Fourier series and integrals" by H. Dym and H.P. McKean, Academic Press, 1985.
"Fourier analysis" by T.W. Korner, Cambridge Univ. Press, 1988.
"Introduction to Fourier analysis on Euclidean spaces" by E.M. Stein and G. Weiss, Princeton Univ. Press, 1971.

For more about special functions:

"Special functions and their applications" by N.N. Lebedev, Dover Publications, 1972.

For a rigorous account of microlocal analysis and pseudodifferential operators:

"Pseudodifferential operators" by M.E. Taylor, Princeton Univ. Press, 1981.

Expository articles on wavelets and quasicrystals:

"How to make wavelets" by R.S. Strichartz, Amer. Math. Monthly 100 (1993), 539-556.
"Quasicrystals: the view from Les Houches" by M. Senechal and J. Taylor, Math. Intelligencer 12(2) 1990, 54-64.

For a complete account of wavelets:

"Ten Lectures on Wavelets" by I. Daubechies, SIAM, 1992.

Index

$\mathcal{S}'$, 46
ψDO, 189
θ-functions, 128
$\mathcal{D}'$, 4
$\mathcal{D}$, 4
$\mathcal{E}$, 81
$\mathcal{E}'$, 81
$\mathcal{F}$, 31
$\mathcal{S}$, 32
L^p norm, 164
x-ray diffraction, 128

a priori estimate, 216
a priori estimates, 183, 192, 200
absolutely continuous, 131
adjoint, 190
adjoint identities, 18, 47
analytic, 112, 120
analytic continuation, 51
analytic functions, 51
anisotripic wave operator, 218
annihilation operator, 141
approximate, 14
approximate identity, 40, 99
approximate identity theorem, 40
approximation by test functions, 98
asymptotic expansion, 149, 188
Atiyah-Singer Index Theorem, 193
atlas, 107
average, 1
average of Gaussians, 52

ball, 3

Banach spaces, 173
Beckner's inequality, 120
Bessel functions, 144
Bessel's differential equation, 161
bicharacteristic, 212
bicharacteristics, 212
Bochner's theorem, 132
Bochner's Theorem, 133
bottom of the spectrum, 142
boundary conditions, 65, 192, 194
boundary value problems, 192
boundedness, 92, 94

calculus of residues, 41
Cantor function, 111
Cantor measure, 91, 111
Cantor set, 90, 111
Cauchy data, 67, 195, 214
Cauchy integral theorem, 42
Cauchy problem, 195, 212
Cauchy-Kovalevska Theorem, 195
Cauchy-Riemann equations, 74, 121, 181
Cauchy-Riemann operator, 178, 192
Cauchy-Schwartz inequality, 137
chain rule, 193
change of variable, 190, 207
characteristic, 209
characteristics, 193
classical solution, 22
classical symbol, 217
classical symbols, 188
coefficients, 28

comb, 126
commutation identities, 142
commutator, 137
compact set, 78
compact support, 78
complementary pairs, 138
complete system, 143
completeness, 152
completion, 82
composition, 189
cone, 197
conjugate harmonic functions, 74
conservation of energy, 67
conservation of momentum, 73
constant coefficient operators, 177,
        186
continuity of, 91
contour integral, 42
convolution, 34, 40, 99
convolutions, 35, 55
cotangent, 208
cotangent bundle, 208
creation and annihilation opera-
        tors, 141
crystal, 128
cut-off function, 99

Daubechies wavelets, 154
decay at infinity, 31, 54
decrease at infinity, 38
degree, 108
derivative of a convolution, 57
diagonal matrix, 144
diagonalizes, 144
differentiability, 38
differential equations, 22, 35
differentiation, 33, 34
dilation equation, 155
dilations, 25
Dirac δ-function, 5
Dirac delta-function, 15
directional derivatives, 193

Dirichlet boundary conditions, 75
Dirichlet problem, 62
distribution theory, v, 220
distributions, 4, 5, 8, 12, 14, 46,
        205
distributions of compact support,
        81
distributions on spheres, 104
distributions with point support,
        85
doctrine of microlocal myopia, 179,
        189
dual cone, 197
dual lattice, 128
duality, 176
Duhamel's integral, 75, 76

eigenfunction, 140
eigenvalue, 140
eigenvalues, 44
elliptic, 177, 187
elliptic operator, 194
elliptic partial differential equations,
        177
energy, 67
entire functions, 121
equipartition of energy, 116
equivalent, 13, 189
equivalent norm, 174
Euler, 157
even, 26
expectation, 131
exponential type, 121
extension, 103, 105

finite order, 85
finite propagation speed, 76, 79
finite speed of propagation, 158,
        199
focusing of singularities, 71
forward light cone, 197
Fourier analysis, 220

Fourier cosine formula, 145
Fourier integrals, 28
Fourier inversion formula, 30, 31,
    38, 49, 57
Fourier series, 28, 66
Fourier sine transform, 146
Fourier transform, 30
Fourier transform of a Gaussian,
    41
Fourier transform of a tempered
    distribution, 47, 48
Fourier transforms, vi, 220
Francois Viete, 157
free Schrödinger equation, 72
freeze the coefficients, 187
full symbol, 178, 187
functional analysis, 173
fundamental domain, 129
fundamental solution, 179, 184, 192,
    196
fundamental solutions, 60

Gabor transform, 139
Gaussian, 39
generalized functions, 1, 2
generating function identity, 160
groundstate, 142

Haar function, 150
Haar functions, 150
Haar series expansion, 151
harmonic, 24, 60, 179
harmonic oscillator, 141
Hausdorff-Young inequality, 120
heat equation, 43, 64
heat operator, 216
Heaviside function, 10, 83
Heisenberg uncertainty principle,
    135
Henri Lebesgue, v
Hermite expansion, 143
Hermite functions, 140, 143

Hermite operator, 218
Hermite polynomial, 143
Hermitian, 132
Hermitian operators, 135
Hilbert space, 18, 173
Hilbert transform, 58, 217
homogeneous, 25, 58, 108
Huyghens' principle, 71, 76, 79
hyperbolic, 196
hyperbolic differential equations,
    126
hyperbolic equations, 196
Hyperbolic operators, 193
hypersurface, 194
hypoellipticity, 181, 192
Hölder conditions, 175
Hölder continuity, 171

index, 192
infinite order, 85
infinite product, 156
inhomogeneous heat equation, 75
initial conditions, 67
integrability, 164
integrable, 13, 114
integral, 110
integral operator, 180, 187
interpolation, 120
inverse Fourier transform, 31

Jacobian, 207

kernel, 35
kinetic energy, 67, 116
Klein-Gordon equation, 75, 158

ladder of eigenfunctions, 142
Laplace equation, 8, 23, 60, 62
Laplacian, 8, 177, 192, 193
lattice, 127
Laurent Schwartz, v, 8
Lebesgue, 173
Lebesgue integration theory, v, 12

Leibniz' formula, 32
light cone, 197
light rays, 213
limits of a sequence of distribu-
    tions, 98
linear functionals, 4
linear operator, 14
linear transform, 14
linearity, 2
Liouville's theorem, 62
Lipschitz condition, 172
local coordinate system, 106
local operator, 211
local theory, 102
localization, 150
locally integrable, 13
location of singularities, 180
logarithmic potential, 60

manifold, 107
manifolds, 192
mathematical physics, vi
maximum speed of propagation,
    71, 124
Maxwell's equations, 71, 76
mean-square, 30
measure theory, v
method of descent, 70
microlocal hypoellipticity, 211
microlocal analysis, 179, 220
microlocal operator, 211
microlocal parametrix, 211
microlocal smoothness, 203
moments, 44
momentum, 73, 136
Monte Carlo approximation, 131
multi-index notation, 83
multiplication, 19, 33, 34
multiplication of, 205
multiplicity, 142

Neumann boundary conditions, 75

Neumann problem, 192
Newtonian potential, 61
noncharacteristic, 193, 195
nonnegative, 14, 89
nonnegative definite, 132
norm, 92, 164
normal direction, 195, 202
null bicharacteristics, 213

observables, 135
odd, 26
open set, 3
operation, 14
operations, 47
operations on, 14
order, 83, 84, 177
order of a distribution, 84
orthogonality, 151
orthonormal family, 151
orthonormal system, 143

Paley-Wiener theorems, 120
parametrix, 179, 187, 192, 211
Parseval identity, 143
Parseval's identity, 28
partial differential equations, vi
partial Fourier transform, 63
partition of unity, 85
periodic, 29
periodic boundary conditions, 65
periodization, 126
perturbation series, 180
ping-pong table, 36
pitch, 139
Plancherel formula, 30, 31, 48
Planck's constant, 72, 136
Poisson integral formula, 64
Poisson summation formula, 126,
    144
polar coordinates, 105
positive, 89
positive definite, 132

positive definite function, 132
positive distribution, 131
positive measures, 90
potential, 9
potential energy, 67, 116
potentials, 60
principal type, 212
probability measure, 130
product, 19
product rule, 184
progagation of singularities, 213
proper, 197
pseudodifferential operator, 188
Pseudodifferential operators, 186
pseudodifferential operators, 220
Pseudolocal property, 190

quantum mechanics, 72, 135, 144
quantum theory, 72, 135
quasicrystals, 129, 220

radial, 26, 44
radial function, 144
rapidly decreasing, 31
real analytic, 195
recursion relation, 160
recursion relations, 149
restriction, 102, 105, 200
restrictions of, 209
Riemann, v
Riemann-Lebesgue lemma, 114
Riemannian metric, 209
Riesz representation theorem, 90
Riesz transforms, 217
rotation, 26, 44

sampling, 140
scaling function, 154
scaling identity, 155
Schrödinger's equation, 43
Schrödinger's equation, 72
Schrödinger equation, 144
Schwartz class, 32

self-adjoint, 141
signal processing, 139
singular support, 180, 205
singularities, 179, 210
smooth manifolds, 104
smooth surfaces, 104
smoothing process, 56
smoothness, 31, 54
Sobolev embedding theorem, 165
Sobolev embedding theorems, 174
Sobolev inequalities, 163
Sobolev space, 190
Sobolev space norm, 173
Sobolev spaces, 173
Sobolev theory, 163
spacelike, 198
special functions, 220
spectral theory, 141
spectrum, 142
sphere, 104
spherical coordinates, 105
spherical harmonics, 88
strictly hyperbolic, 199
strip-projection, 129
structure theorem, 96
structure theorem for $\mathcal{D}'$, 84
structure theorem for $\mathcal{E}'$, 83
structure theorem for $\mathcal{S}'$, 84
structure theorems, 82
summability, 39
sup-norm, 92
support, 78, 180
surface integrals, 68
symbol, 178
Symbolic completeness, 189

tangential direction, 202
tangential directions, 195
Taylor expansion, 86
temperature, 1, 64
tempered, 8, 46
test functions, 1, 2

top-order symbol, 178, 187
translation, 33
translations of, 15
traveling wave, 22
triangle inequality, 164

uniformly elliptic, 187

vanishes at infinity, 114
variable coefficient operators, 186
variance, 135
vibrating string equation, 8, 9
vibrating strings equation, 22

wave equation, 67, 116, 196, 197,
        213
wave front set, 202, 203
wave function, 72, 135
wave operator, 200, 213
wavelet, 154
wavelets, 150, 220
weak solution, 22
well-posed, 196

Zygmund class, 172, 215

www.ingramcontent.com/pod-product-compliance
Lightning Source LLC
Chambersburg PA
CBHW060254220326
41598CB00027B/4104